Choosing Ecological Sewage Treatment

Nick Grant, Mark Moodie, Chris Weedon

©Nick Grant, Mark Moodie and Chris Weedon, August 2012

The Centre for Alternative Technology
Machynlleth, Powys
SY20 9AZ, UK
Tel. 01654 705980
pubs@cat.org.uk • www.cat.org.uk

ISBN 978-1-902175-78-2
1 2 3 4 5 6 7 8 9

The right of the authors to be identified as the author of this work has been asserted by them in accordance with the Copyright, Designs and Patents Act 2005.

All rights reserved. No part of this publication may be reproduced, stored in a retrieval system, or transmitted, in any form, or by means, electronic, mechanical, photocopying, recording or otherwise, without the prior permission of the copyright owner.

The details are provided in good faith and believed to be correct at the time of writing. However, no responsibility is taken for any errors. Our publications are updated regularly; please let us know of any amendments or additions that you think may be useful for future editions.

Editor: Allan Shepherd
Assistance: Peter Harper, Graham Preston, Lesley Bradnam and Jules Lesniewski
Typesetting and layout: Graham Preston
Illustrations: Graham Preston, Maritsa Kelly and the authors
Cover design: Annika Faircloth

Published by CAT Publications, CAT Charity Ltd.
Registered charity no. 265239.

Mixed Sources
Product group from well-managed forests and recycled wood or fiber
www.fsc.org Cert no. TT-COC-2200
© 1996 Forest Stewardship Council

Printed and bound by CPI Group (UK) Ltd, Croydon, CR0 4YY

"The society which scorns excellence in plumbing because it is a humble activity and tolerates shoddiness in philosophy because it is an exalted activity, will have neither good plumbing nor good philosophy: neither its pipes nor its theories will hold water."

John W. Gardner

The authors

Nick Grant
Nick trained as a design engineer and, apart from a spell at CAT in the 1980s, has been self-employed for most of his adult life. He met Mark while living in a community in Dorset, where they built their first reed bed system. They formed Elemental Solutions in 1996 providing design and consultancy in all aspects of wastewater and water efficiency. Nick and his partner Sheila built their own 'eco-house', which reawakened an interest in wider aspects of eco-design and building in particular. Currently most of his work is as an energy consultant specialising in the Passivhaus approach to ultra efficient building. Recent projects include three Passivhaus schools, as well as a wide range of housing from one off selfbuild homes to flats.

Chris Weedon, BSc, ARCS, DIC, PhD
Chris trained as a biochemist and worked at CAT (1992-96) having previously spent four years in R&D as a microbial physiologist in the pharmaceutical industry. Whilst at CAT he monitored and developed the reed bed systems and established the sewage treatment consultancy. He is now the director of Watercourse Systems Ltd, designers and installers of aquatic plant wastewater treatment systems. In 1997 he developed the Compact Vertical Flow reed bed, now widely used in Europe, with over 60 around the UK, one of which serves his family home in Somerset.

Mark Moodie
Mark spent six years training in medicine and homoeopathy before graduating to sewage in 1990. He has recently moved with his family on to the main sewer giving him, in theory, more time to devote to understanding the connection between psyche and soma.

Contents

Foreword	1
Introduction	3
The subject of this book	3
A note on vocabulary	4
A warning about pathogens	4
Chapter One: The Big Circle	7
The big circle	7
Sewage and the big circle	7
Sewage breakdown and the mineral solution	8
Conditions for digestion	9
Reintegration	10
Using the big circle	11
Summary	12
Chapter Two: Treating Sewage	13
Why treat sewage?	13
What is in domestic sewage	14
Aerobic or anaerobic?	15
Treatment stages	17
Preliminary treatment	17
Primary treatment	17
Secondary treatment	17
Nitrification	18
Tertiary treatment	19
Summary	20
Chapter Three: Sewage Treatment Systems	21
Choosing a treatment system	21
Sewage treatment technologies	22
Cesspools	23
Septic tanks	24
Solids separators	26

Percolating filters	28
Package plants	30
Vertical flow reed beds	36
Subsurface horizontal flow reed beds	39
Free water surface flow reed beds	42
Ponds	42
Leachfields and soakaways	44
Willows and trenches	46
Living machines	48
Living soakaways	49
If it ain't broke don't fix it	51
Summary	52
Chapter Four: Monitoring and Regulations	53
Monitoring methods	53
The jam-jar method	54
The turbidity tube	54
Biochemical tests	55
The major determinands: SS and BOD	55
Other determinands	59
A worked example	65
Biotic indices	68
Regulators and regulations	72
Environmental permits	73
Environmental Health Department	74
Summary	75
Chapter Five: Avoiding Generation of Blackwater	77
Soil versus water	77
Humus	78
How to make humanure	80
Composting toilets	80
Composting toilets and health	82
The challenges of composting human muck	83
Urine collection	86
Electric toilets and chemical toilets	88
Returning humanure to the soil; a word of caution	90
Designer's flowchart	92
Designer's checklist	94
Summary	96
Chapter Six: Using Domestic Water Wisely	97
What you can put down the drain	97

Detergents and household cleaning agents	98
Eco balls?	100
Other chemicals	101
Water conservation in the home	102
Why conserve water?	102
Reasons to save water	102
Order of priorities: reduce, reuse, recycle	103
Developing awareness, changing habits	104
Domestic water audit	104
Changing habits	104
Leak plugging	105
Technical solutions	105
Toilets	106
Showers, baths and basins	109
Dead legs	110
Combi boilers	110
Taps	111
White goods	112
Water meters	113
Gardens	113
Reusing water	114
Rainwater harvesting	114
Water recycling	115
Summary	117
Chapter Seven: Back to Life	119
Humus	120
The nutrient solution	120
Discharge to water	120
Discharge to soil	121
Aquaculture	122
Irrigating willows	124
Irrigating other plants	125
Last Word	127
Case Studies	129
Glossary	157
Resources	163
Index	165

Foreword

In my own work with the National Trust, I find that a good 75% of property management discussions revolve around the correct treatment of sewage.

And so they should — as a conservation issue the disposal of human waste is the most all-pervading cause of environmental health problems across the globe. As human populations expand, it is inevitable that the water each of us consumes or the land each of us uses will be affected by the effluent from all those people living 'upstream'. As our knowledge of the cyclic nature of the environment develops (with an increasing sense of urgency), we realise that the people who live 'downstream' of us also affect the health of our environment: through the sewage sludge that has to be disposed of; through the demand for water leading to more reservoirs and pipelines; and more insidiously through the acidification of lakes and streams from the deposition — in rainfall — of nitrogen derived from emissions from sewage disposal many miles away...

We have to be much cleverer in the way we regard human waste if we are going to solve some of the critical environmental problems of our time. We shall certainly have to invest in more water-efficient sewerage systems; we shall have to develop sewage treatment as a recycling process rather than a waste disposal process; we shall have to be much more selective about the kinds of waste which are put into sewage systems; above all, we have to change the all-too-common public attitude to waste - out of sight, out of mind (i.e. out of my site, out of my mind).

Thankfully, there is hope — a dedicated band of 'sewer-ers' is researching and applying new technologies to sewage treatment. This includes the three authors of this book and CAT, but there are many others in commercial companies, water utilities, intermediate technology groups and universities who are persistently pushing new ideas and solutions into practice. The National Trust itself has taken a lead and installed dry compost lavatories, constructed wetland treatment systems, waterless urinals, 'Green cones' and improved septic tank systems at many of its properties. The benefits of these are apparent: a higher standard of water environment; reduced consumption of precious natural resources; and especially a greater awareness of the sensitivity of 'human effluent recycling systems' to damage from detergents, bactericides and inappropriate waste such as oils and grease and personal sanitary wear. These systems enable nutrients to be recycled as well as humus and water, so that a 'virtuous circle' of food production to healthy eating to return of natural resources to the environment can be established.

To all those who pick up this book to browse, YOU are the solution: 'bon appetit' and 'set to'.

Rob Jarman
Sustainability Director, National Trust

Introduction

The Subject of this book

Our main aim in writing this book is to empower and enable you to understand, and be a good guardian of, your own sewage treatment system.

During our daily work as designers of natural sewage treatment systems many people have approached us saying, 'My septic tank is broken, I need a reed bed or pond system.' Often a little information and orientation is all that is required to enable an old system to be brought back to its former glory. If you are seeking to revive an existing system then we hope that *Choosing Ecological Sewage Treatment* will provide those insights.

Other people have moved to the country and for the first time are faced with the unglamorous prospect of dealing with all the stuff they put down their drains: 'What do I do?' The book is also intended as a first step for those who need a new system but who do not know what technological options are available, or what it would be useful to understand in order to make a sensible choice between the different possibilities.

Our approach is first to provide a contextual framework for understanding the processes of sewage treatment. Assuming that you will be using a flush toilet (as opposed to other options), we move on to describe the basic principles of sewage treatment and the options for non-mains treatment, some of which are regarded as 'green' technologies and others that are regarded as conventional.

This leads to a discussion of how to determine the extent to which your attempt at sewage treatment has been successful.

Then we go back a step to question the necessity of using water to transport human muck, and discuss the alternatives (such as compost toilets) and their associated benefits and risks. In a similar vein, we then deal with appropriate use of water in the home, which in itself can play a major role in determining your treatment system's design.

Finally we discuss the route by which treated sewage can be re-introduced into the environment.

A note on vocabulary

The question of appropriate terminology has been considered. Should we be twee and tabloid, or blunt and widely understood? Perhaps we ought to stick to the terms of the scientific establishment and risk losing some folk to avoid offending others. But we would still have problems. For instance, the 'proper' term for a single coherent aggregate of faeces is not a 'faece', but 'bolus' (a synonym for 'turd'); but does anyone out there actually use it?

There is no intention to offend with our choice of words. We have tried to be varied.

'Wastewater' is also a difficult word for us. In the absence of a more convenient term, we have used it, while being fully aware of the potential value of 'waste' water and its contents.

One term you may not be familiar with is 'humanure', intended to emphasise that properly treated human muck has potential for growing plants. When it comes down to it, shit by any other name will smell as sweet.

On a slightly different note, there is often confusion between 'sewage' and 'sewerage'. 'Sewage' is the stuff that is transported and treated by 'sewerage'. Sewerage is the term for the pipes, manholes, pumping stations and other hardware, which enable all that sewage to get to its destination and be treated. Municipal scale sewerage is collectively known as 'the mains'.

Other specialised terms are discussed in the text and many are collected in the glossary.

A warning about pathogens

An organism is said to be pathogenic if it is implicated in a disease. If the disease is in a person, then we are dealing with a human pathogen. The faeces of healthy individuals contain vast numbers of bacteria, the majority of which are not pathogenic. However, a small proportion of the bacteria could cause stomach upset if ingested. Moreover, the excrement of people suffering certain diseases contains billions of pathogens. These include viruses (such as hepatitis A and polio virus), bacteria (such as those implicated in cholera and typhoid fever) and larger organisms (such as the protozoan associated with amoebic dysentery, or the flat-worm involved in river-blindness).

Due mainly to the improvements in municipal sanitation, serious waterborne diseases are extremely uncommon in the UK. A few pathogens, including those associated with river-blindness as well as Weil's disease (spread in rat urine), can enter the body through intact skin! However, the main danger comes from ingestion of sewage. Ingestion of sewage-contaminated water can lead to stomach complaints, or worse if a kidney or other vital organ becomes infected.

The more intimate your contact with sewage the more chance you have of getting sick. Wear gloves and overalls if you have to work with sewage, and a face-mask to avoid splashing your mouth. After working with sewage avoid the kitchen and eating areas until you have washed thoroughly... all basic common sense, but please take heed. Santé.

Chapter One
The Big Circle

Before exploring the details and practicalities of choosing a sewage system we will take a moment to establish the general context within which all sewage treatment occurs. Such a framework can be used to understand different sewage systems and evaluate their strengths and shortcomings. The more one understands the rationale behind sewage treatment, the more creative one can be in devising sewage solutions.

The Big Circle

Sewage treatment is a specific case of a more general and ancient process. Inanimate matter (such as minerals, gases, water) is constantly being incorporated into organisms and so taken out of equilibrium with its environment. Upon death, the organisms' bodies undergo a process of decay, returning to equilibrium with the inorganic world.

So for our purposes we can distinguish three phases of matter. In the first it is mineral or inorganic. When incorporated in the living tissues we can say matter is part of an organism. And when the organism dies we can say that it is organic matter until, once the process of decay has had its way, it is once again indistinguishable from the rest of the inorganic world.

This recycling of material is central to sustaining life and whether we are farmers, gardeners, or householders we are involved in this cycle, which we call the 'Big Circle' (figure1.1).

Sewage and the Big Circle

What is sewage? To be very general we can say domestic sewage is a mixture of water and the various types of organic matter that we send through the plumbing; faeces and urine, food scraps, hair, and toilet paper for example. Domestic sewage also contains household chemicals and detergents (figures 1.2 & 1.3), and we'll discuss these in subsequent chapters.

For around 96% of UK residents, all this stuff will disappear down various

8 Choosing ecological sewage treatment

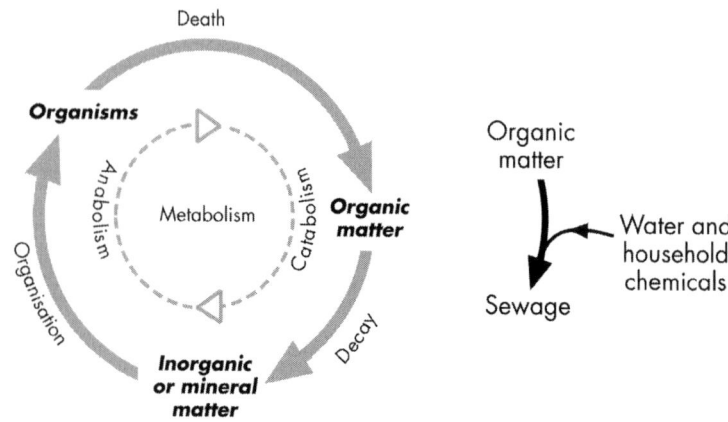

Figure 1.1. The big circle (left) and 1.2 sewage (right).

loos and plugholes, never to be seen by them again. These folks are 'on the mains', and the majority are required by law to connect to the municipal infrastructure. For historical reasons this makes a lot of sense, and we're not suggesting that anyone break free from the mains.

But for those not on the mains and who want (or are compelled) to do 'the right thing' with their wastewater, what happens after the flush needs to be considered too. That's what this book is all about.

Sewage breakdown and the mineral solution

Once the sewage disappears down those loos and plugholes, as with all organic matter, it breaks down. The complex organisation that was, for example, a cabbage undergoes a series of processes that make it less and less recognisable as cabbage and more and more akin to the breakdown products of other organic bits of sewage, for instance, former carrots or chocolates.

At all stages of breakdown of our cabbage some 'energy' is released. We get energy when we eat the cabbage. When we excrete it our wastes become a feast, a source of energy, for other tiny creatures. These microorganisms will go to work, breaking the organic matter down into smaller and simpler fragments.

The breakdown occurs only as far and as fast as conditions allow. If these are right, the creatures will continue to break down our wastes, releasing various gaseous products. Eventually all that is left is a solution of minerals dissolved in water.

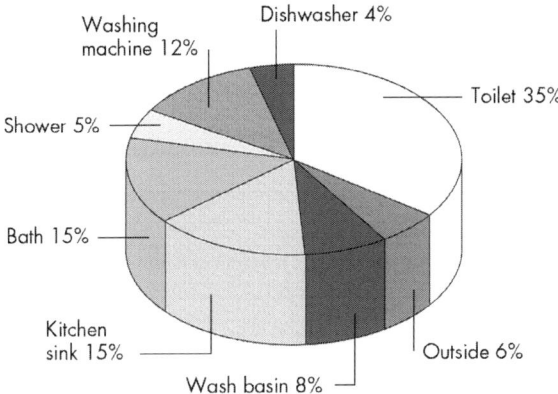

Figure 1.3. Water usage in a typical household.
Source: Anglian Water Survey of Domestic Consumption.

The process of breaking down organic waste is often called 'mineralisation' (figure 1.4). The process is also known as digestion, catabolism, disintegration, or dissolution and is desirable in a sewage treatment system. In the next chapter we'll say why this is so.

Conditions for digestion

What are the conditions necessary to turn the sewage into a mineral solution as efficiently and completely as possible? For the sewage treatment systems covered in this book, there are three main conditions that must be satisfied: sufficient oxygen, the right temperature range, and somewhere for the microorganisms to live and grow. Insufficient oxygen would lead to much slower decomposition and the generation of toxic and malodorous breakdown products (see chapter two). As the temperature drops, biological and chemical activity slows down considerably. We do not suggest heating sewage to improve treatment but if a high quality effluent (for example low ammonia levels) is required in a cold climate, then consideration will need to be given to avoiding unnecessary heat loss from the usually warm sewage leaving the house. The final important consideration is habitat for the microorganisms. Simply aerating warm sewage as it passes through a tank will not lead to effective treatment, as the growing organisms will be washed out before they have reached their optimum growth phase. Typically, sewage systems use some sort of matrix such as stone, sand, fibre, clinker or moulded plastic for the bugs to grow in. Alternatively the bugs are encouraged to form

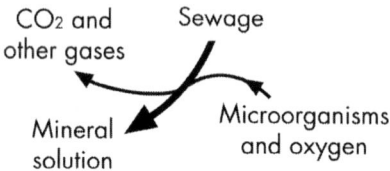

Figure 1.4. Sewage digestion.

large colonies that are suspended in the sewage that is being treated – for example in activated sludge systems and treatment ponds (see chapter three). That's the basic equation of sewage treatment.

We should address a common misunderstanding about what sewage treatment achieves; although treatment of sewage includes removing the contaminating material from the water, the primary process is one of transformation. Apart from the greater part of the carbon (which leaves as carbon dioxide gas), the mineral solution at the end will contain a great deal of the same atoms of stuff as the sewage did before the bugs started breaking it down. (It is, at a later stage in treatment, possible to remove some of these minerals from the solution. See chapters two and seven.)

Reintegration

The story does not end here though. Catabolism is only part of the cycle of metabolism — the Big Circle. The other half is anabolism — the construction of organisms from simple minerals (see figure 1.1).

Let's reconsider the mineral state. We stated that the minerals are stable because they will not break down into smaller fragments. But their environment is full of potential, one in which creatures can multiply and grow.

Best adapted to using the raw components of the mineral solution are plants and microorganisms. They find it very nutritious and for this reason the output of a sewage treatment system is sometimes called a nutrient solution. Microorganisms have many methods of getting what they need for energy and growth. These have been used to categorise bacteria into different groups (box 1.1).

The first organisms to join the feast will be relatively simple autotrophs. These will become food for the first heterotrophic organisms, which in turn become food for further heterotrophs. Holding such a solution before the mind's eye, one would see phytoplankton and zooplankton. Then would appear more complex organisms: micro-invertebrates, larger plants, insects

Box 1.1. What-o-trophs?

Bacteria need both an energy source and a carbon source. To produce their energy they can use light (in which case they are called 'phototrophs') or chemical compounds ('chemotrophs'). To build their cell material the carbon can come from either organic material ('heterotrophs') or CO_2 ('autotrophs'). Combinations of all four groups exist, giving a rich variety of metabolic types.

Category	Energy source	Carbon source	Examples
Photoautotroph	light	CO_2	photosynthetic bacteria (all green plants)
Photoheterotroph	light	organic C	some purple and green bacteria
Chemoautotroph	chemicals	CO_2	nitrifying bacteria
Chemoheterotroph	chemicals	organic C	most bacteria, fungi, animals

and their larvae, amphibia and fish. Without imaginative restraint would appear organisms that can live around the aquatic solution: terrestrial mammals and birds, and even organisations who are so complex they have complexes — us.

We can call this ascending limb creation, organisation, integration, or anabolism (figure 1.5).

Using the Big Circle

The map is not the same as the territory. For example, the downward half of the cycle is not always completed by each carcass before it is swept up into a new organism. Were we to allow for this and all other variations on the theme we would have to employ animation, and something like a writhing bowl of spaghetti would appear, rather than an orderly static circle.

Nevertheless the Circle enables us to see what we might manipulate, what it would be foolish to change, and what it would be foolish to leave alone.

For instance, we might ask ourselves at what point in the cycle we should release our treated sewage back to the wider world in a particular situation. We can ask ourselves why we added the water to the organic matter to begin with and, assuming we decided that this was an appropriate thing to do, what would be a sensible amount and quality of water. Questions of finance, understanding, level of interest and taboo can be inserted at various points and the map changes about its basic form. Feel free to play.

12 Choosing ecological sewage treatment

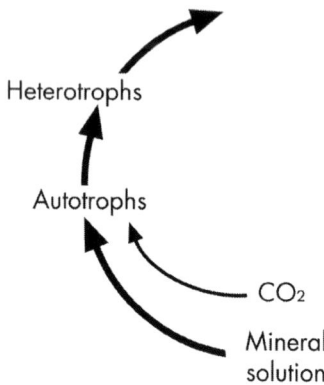

Figure 1.5. The ascending limb.

Summary

Organisms are in a process of continuous transformation, which we have illustrated as the Big Circle.

Since, in the case of most humans in the British Isles, expulsion of bodily wastes is almost exclusively to water, this complicates the picture by creating sewage. Hence we have presented you with the 'Big Sewage Circle' (figure 1.6), a model in which general principles of sewage treatment and its relationship to the wider environment can be illustrated.

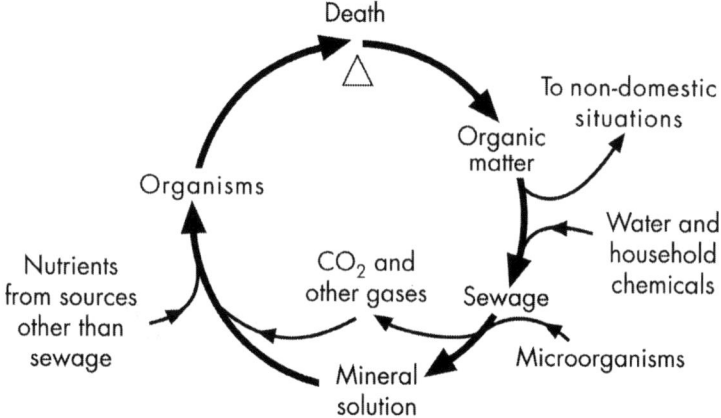

Figure 1.6. The big sewage circle.

Chapter Two
Treating Sewage

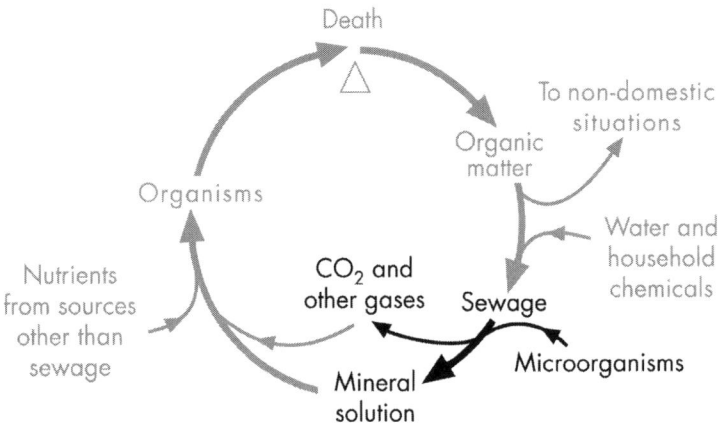

The Big Circle set the scene for our discussion of sewage. We have seen that the same material cycles round and round. On some cycles this matter is added to water by humans, producing sewage.

This chapter deals with sewage treatment, the processes by which sewage is transformed, thus cleaning the water.

Why treat sewage?
We stated in the previous chapter that microorganisms break down the organic matter in sewage. Those microorganisms that require oxygen for their respiration are extremely effective at getting dissolved oxygen if there is any available. In fact, they are much more effective than the more highly evolved creatures that live in unpolluted streams, such as fish.

This brings us to the first reason for treating sewage. If you ran a pipe from your house to the nearest stream and dumped your sewage directly into it, the microorganisms would use up all the oxygen dissolved in the stream

Box 2.1. What is in domestic sewage?

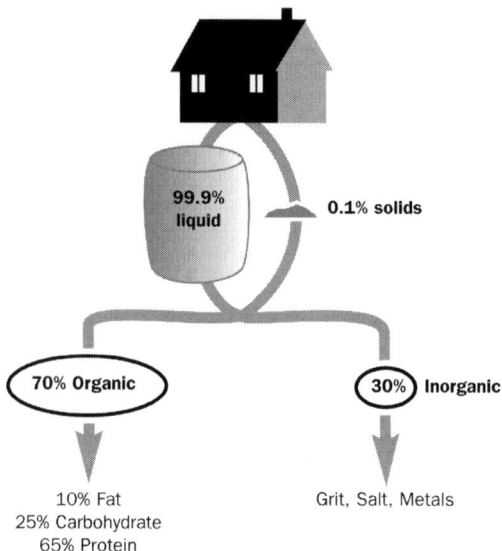

Everything that people put down their toilets and drains from sinks, washing machines, etc., becomes sewage. Unfortunately, some people mistakenly treat their drains as another dustbin! Sewage includes:

Screenable solids (big lumps of material that can be removed from the water by a coarse screen): paper (lavatory paper, paper towels, etc.), food particles, tampons, sanitary towels, condoms, nappy liners, incontinency pads and bags, a myriad other non-biodegradable sanitary items (ear-buds, plastic wrappers, etc.) tablets, other medicines and a seemingly endless list.

Non-screenable solids (tiny particles, most of which remain suspended indefinitely in water and which give it a cloudy appearance): bacteria, faecal particles, food particles, fats, oils and greases, detergents and soaps, washings (from clothes, skin from bodies, etc.), sediment.

Dissolved material (which may contribute colour but not cloudiness): **organic matter**: proteins, urea (from urine), carbohydrates, fatty acids, other organic molecules in small amounts (DNA, etc.); ions of ammonia, chloride, nitrate and nitrite, phosphate, sulphide; oxygen, minerals, metals and trace elements, other molecules in small amounts (vitamins, etc.).

water as they broke down that sewage.

The organic matter would get broken down but there would be no oxygen for fish. To protect the fish, and various other aquatic organisms that depend on having oxygen in the water, the organic matter in your sewage must be digested before it reaches your local stream.

Because organic matter uses up dissolved oxygen when decomposing, a measure of the amount of organic matter in a body of water can be ascertained by recording how much oxygen is being removed (in other words demanded by the microorganisms present) from the water. This measurement is known as the biochemical oxygen demand (BOD).

Another readily testable measurement for the amount of matter in sewage is to see how much is caught in a filter. This is called the 'suspended' solids (SS). SS and BOD are the two qualities of sewage most commonly measured by the authorities.

Ammonia (strictly a mixture of dissolved ammonia gas, NH_3, and ammonium ions, NH_4+) is perhaps the next most important substance to monitor. Many freshwater organisms find even low concentrations of ammonia toxic. It can be produced in large amounts in domestic sewage, and like organic matter, demands oxygen for its breakdown.

Another reason for sewage treatment is that sewage contains potentially pathogenic materials and organisms: bacteria, viruses, worms and chemical compounds. The need for effective sewage treatment is clear when one considers that near the mouth of some rivers, people are drinking water that has previously quenched the thirst of many others, cycling through their bodies via sewage works to the river, and back again!.

Aerobic or anaerobic?

To reiterate the aims: the first step is to transform the organic matter into the mineral constituents; the next step is to remove these from the water. To achieve these goals we must create conditions in which the organic matter can be digested by microorganisms: sufficient moisture, warmth and air. Moisture is not short in a water-borne sewage system. Warmth is sometimes lacking, especially in winter but in the UK climate the temperature within a treatment system rarely becomes low enough to completely halt degradation.

Oxygen, unlike the other factors, can easily be in short supply and careful provision must be made to ensure sewage has good access to it. Degradation can occur in the absence of oxygen (anaerobic conditions) but the metabolic

Box 2.2. Metabolism in sewage treatment

End products of aerobic and anaerobic decomposition (simplified)

Element	Aerobic	Anaerobic
carbon C	carbon dioxide CO_2	methane CH_4
hydrogen H	water H_2O	hydrogen H_2
oxygen O	oxygen O_2	water H_2O
nitrogen N	nitrate NO_3^-	ammonia NH_3
phosphorus P	phosphate PO_4^{3-}	phosphine PH_3
sulphur S	sulphate SO_4^{2-}	hydrogen sulphide H_2S

processes involved are much slower than in the presence of oxygen (aerobic conditions). The range of metabolic processes is complicated but we can generalise by saying that in aerobic conditions the chemical elements will emerge from the degradation process in combination with oxygen; in anaerobic conditions they will emerge combined with hydrogen (Box 2.2).

The products of aerobic decomposition are odourless, non-toxic and water-soluble. They represent full breakdown of the various contaminants of sewage water and are, therefore, stable. They are the constituents of the mineral solution.

Some products of anaerobic degradation, such as alcohols and fatty acids are more chemically complicated than those shown in Box 2.1. Many are toxic to organisms adapted to conditions in an oxygen-rich environment. Some are noxious gases like hydrogen sulphide and volatile fatty acids. Some of the gases are potentially explosive. A notable feature of most anaerobic products is that they are smelly. One's nose knows when there is insufficient oxygen available for the aerobic breakdown of sewage.

The whole anaerobic gas mixture is sometimes referred to as 'biogas'. Anaerobic digestion is efficient only at quite high temperatures (typically around 35°C). In warmer climates anaerobic digesters are commonly used to break down sewage sludge although in the UK they are economically viable only on a large scale (for example populations greater than 5,000). Anaerobic breakdown of organic matter is common in nature, occurring in many wetlands, for instance.

Treatment stages
Sewage treatment systems are characterised by the level of treatment they provide.

Preliminary treatment
Preliminary treatment is the physical removal of relatively large or heavy solids (grit, wood, rags and so on). This is usually achieved by passing the incoming sewage through a screen with bars 25-50mm apart (see figure 2.1), or through a net of similar gauge. The methods are exclusively physical and since very little organic contamination has been diverted from the liquid, the BOD is not much reduced.

Preliminary treatment is rarely required for domestic sized systems, unless there are pumps that lift the raw and unsettled sewage.

Primary treatment
Primary treatment usually involves slowing the sewage down – putting the effluent into a chamber so that the suspended solids (mainly organic material) either settle to the bottom of the chamber through the force of gravity, or float to the surface by buoyancy. This process is predominantly physical. Since the material that settles out is organic, the total BOD of the sewage that continues through the system is reduced.

In domestic situations, this settlement chamber is usually a septic tank, which retains around 30–50% of the BOD and SS. The effluent from this stage is still highly polluting and if you were to collect some in a jam-jar you would notice that it is a fairly opaque grey colour with a distinctly sewagey smell (figure 2.2 on page 19, left hand jam-jar).

Secondary treatment
This process removes most of the remaining BOD (mainly soluble organic material) and SS. It is also known as biological treatment since it depends on microorganisms breaking down the organic material in the sewage. Given the right conditions, microorganisms will thrive in the effluent of primary treatment and, in feeding on it, break it down. Biological or secondary treatment involves creating an environment where this can happen at an accelerated rate, in a relatively small space, away from valuable habitats such as trout streams. A sample of effluent from a secondary stage will, if all is well, show that our smelly grey soup has been transformed into a relatively clear liquid.

Figure 2.1. Preliminary screening at the small treatment plant of an army barracks.

If we allow the slightly turbid secondary effluent to stand for a few hours, any remaining solids settle out, leaving the final product of secondary treatment; a clear 'mineral solution'.

In practice we can achieve this settling of the secondary solids by adding a second tank and the result is an effluent that should satisfy the regulatory authorities (figure 2.2, right hand jam-jar).

Whilst the details vary we can summarise the recipe for secondary biological treatment system as:

1. Somewhere for the bugs to live.
2. A source of oxygen.
3. A way of removing the dead microorganisms (settlement tank).

Nitrification

Nitrification is the transformation of ammonia to nitrate. It is achieved by a few specialist bacteria, which must have a good supply of oxygen. Because of this, the concentration of BOD must be sufficiently low, so the nitrifiers are not out-competed by more robust bugs. So nitrification occurs during the latter stages of secondary sewage treatment.

Figure 2.2. Jam-jars of sewage (left) after a septic tank and (right) settled after good secondary treatment (by a vertical flow reed bed).

Tertiary treatment

Tertiary treatment refers to any or all of the following:

- Removal of further remaining BOD, SS and ammonia.
- Removal of nutrients, in other words nitrogen and phosphorus compounds.
- Removal of pathogenic organisms.

Some people use this term to refer specifically to only one of the above changes to the sewage. Generally, though, tertiary treatment is used loosely to refer to any combination of the above.

It is usual for a sewage treatment plant to discharge secondary quality water, leaving further purification of this water to the organisms in the natural environment. Nevertheless, legal standards are rising all the time and there are situations where a higher quality of effluent is required, for example, when a discharge is into a very small stream, or near bathing waters.

Tertiary treatment involves taking the mineral solution through a further biological, physical or chemical step. On a municipal scale, dissolved contaminants can be economically removed by precipitation (the solids being chemically forced out of solution), flocculation and settlement.

A range of existing technologies and possible formats will be covered in some detail in chapter three.

Summary

To minimise the adverse effects of sewage on the environment and on people, it is necessary to degrade the contaminating constituents in the water. This involves removing gross particles (preliminary and primary treatment) then biological transformation of the remaining contaminants into stable, 'mineral' form (secondary treatment). Further removal of contaminants is possible (tertiary treatment) by physical, chemical and/or biological means.

Figure 2.3. Sampling effluent at CAT.

Chapter Three
Sewage Treatment Systems

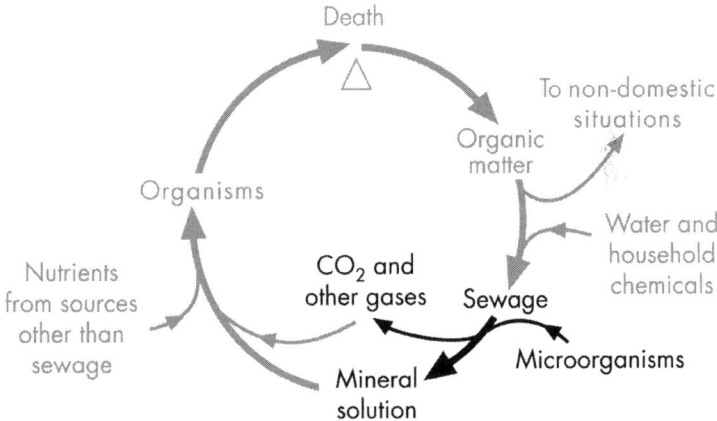

Having looked at the context for sewage treatment and introduced you to the basic processes involved, we now turn our attention to the hardware involved in small scale, on-site treatment of sewage. The chapter is arranged as follows:

1. Choosing a treatment system.
2. Small scale sewage treatment technologies.
3. If it ain't broke, don't fix it.

Choosing a treatment system

To repeat an important point: if you can connect to the mains then do so. Many flagship so-called 'sustainable' building projects insist on trying to disconnect from the mains. This rarely makes sense either ecologically or economically; don't do it.

To guide the transformation of raw sewage into mineral solution in a way that is efficient and acceptable to society means making use of several processes and bits of equipment that satisfy the basic requirements we have sketched out in the preceding chapters.

The design of any good sewage treatment system involves a number of principal factors. A minimum requirement is that the system meets the standards set by the relevant local authorities. After that come various considerations, the importance of which varies with each individual situation:

- Available land area and gradient.
- Level of maintenance you are prepared and able to give.
- Degree of smell and aesthetic impact you (and your neighbours!) will tolerate.
- Ecological sensitivity of the site.
- Amount of money you can spend on it.
- Power requirements and availability.
- Whether you wish to build the system yourself or want to buy one 'off the shelf'.
- Who might use the property next.

A starting place and reference point for choosing the most suitable treatment system is the 'Designer's checklist of system types' and the 'Designer's flowchart' beginning on page 92. The flowchart gives a broad summary of one possible decision-making process and the checklist gives the different characteristics of the various systems and many of the factors we think you will need to consider.

In this chapter we briefly describe each method or technology with the aim of allowing you to compare them and decide which one is for you. Note that each different method is rarely more than part of a whole sewage treatment system. Where individual elements are commonly used together (for example septic tank and leachfield) this is stated. However, almost any combination is possible (although we hope it will be clear that it would be ludicrous, for instance, to have a septic tank receiving the effluent of a solar pond).

Sewage treatment technologies

The following technologies are discussed and are intended to cover the full range of 'wet' on-site approaches available in the UK:

1. Cesspools.
2. Septic tanks.

3. Solids separators (Aquatron).
4. Percolating filters.
5. Package plants.
6. Vertical flow reed beds.
7. Subsurface horizontal flow reed beds.
8. Free water surface flow reed beds.
9. Ponds.
10. Leachfields and soakaways.
11. Land treatment (willows).
12. Living machines.

1. Cesspools (containment)

A cesspool (figure 3.1) is a relatively big tank (18m^3 for one or two people, plus 6.8m^3 per additional person) that has an inlet but no outlet. Cesspools do not treat sewage but simply store it until it is removed by a sludge tanker. Where the ground is unsuitable for accepting discharged effluent and in places where no receiving watercourse is available, they are the only conventional solution.

A typical four-person family produces around 700 litres of sewage a day and an average tanker can hold 8,000–12,000 litres. This means that a tanker load must be removed approximately every two weeks at a cost of around £100–200 a time (depending on distance tankered, locality and so on). Hence the use of water conservation measures, dry toilets, or greywater irrigation, will result in quick payback.

Cesspools are often installed because of high groundwater levels, so any leaks in the tank can mean the tank filling with clean water leaving no room for the sewage. In general, owing to high running costs and potential to cause pollution due to neglect, cess pools are frowned upon by the regulatory authorities and are a *last* resort!

Pros:
- The only conventional solution in some situations.
- Zero effluent discharge to the environment.
- No treatment processes to go wrong.

Cons:
- Transport plus final disposal have high ecological impact.
- Emptying is expensive and costs continue to increase.
- Not permitted for new developments in some areas (for example because of susceptibility to sabotage or neglect, by users trying to avoid emptying charges!).

24 Choosing ecological sewage treatment

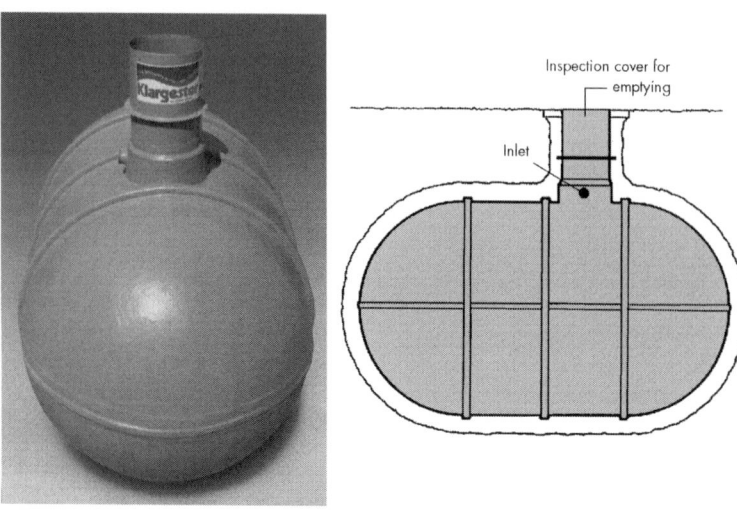

Figure 3.1.a. Photograph of a prefabricated cesspool (not installed) *(courtesy of Klargester Environmental Engineering Ltd.).*
Figure 3.1.b. Side view of a Prefabricated Cesspool *(courtesy of Klargester Environmental Engineering Ltd.).*

2. Septic tanks (primary treatment)

Septic tanks are easily distinguished from cesspools, which have an inlet but no outlet – septic tanks have both. Septic tanks (figure 3.2) are much smaller than cesspools because they retain only the equivalent of around a day's worth of flow. Septic tanks must be sized according to the number of people served but the regulations (BS.6297 and Building Regulations) recommend that a septic tank be no smaller than 2,700 litres. To calculate the required working volume (V) BS.6297 provides the following formula:

V(litres) = (P x 180) + 2,000 where P is the number of people served.

For example a 6 person system: V = (6 x 180) + 2,000 = 3,080 litres.

Tanks with a smaller volume per person are often used for primary settlement in municipal systems where the sludge is regularly decanted. These are 'settlement tanks' (figure 3.3) and are rarely used in private domestic systems.

A septic tank has two chambers (occasionally one or three). Raw sewage enters the first chamber, where most of the solids either settle or float leaving the clearer liquid between to pass out of the tank or on to the next chamber for further settlement. The incoming wastewater displaces the water already

Sewage Treatment Systems 25

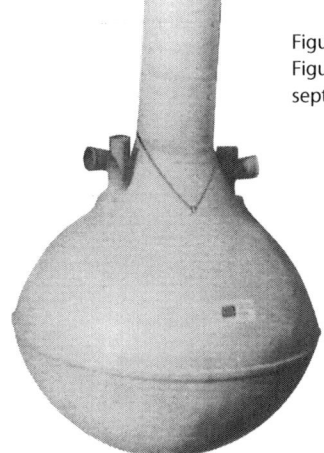

Figure 3.2.a. Prefabricated septic tank (not installed).
Figure 3.2.b. (below) Cross-section of a constructed septic tank.

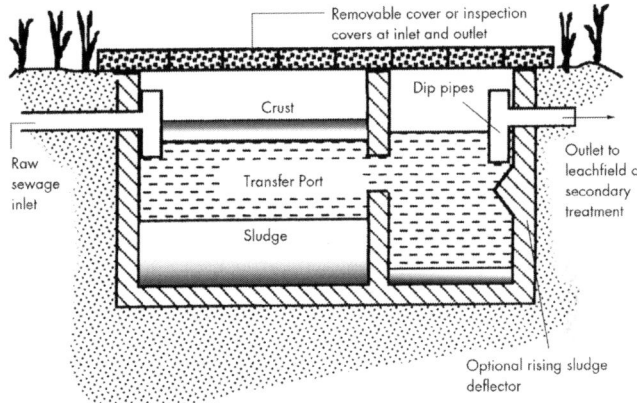

in the tank, much as a bath would overflow were one to leave the tap running. Do not be surprised, therefore, when looking inside, to find it 'full'. Whilst water is displaced from the final chamber by incoming sewage from the home, most of the gross solids and about one third to one half of the organic load is retained in the tank. This solid material must be removed periodically (from every six months to every six years) by sludge tanker.

Septic tanks provide only primary treatment and so must be followed by a leachfield or secondary treatment system. Discharge directly to watercourse is illegal.

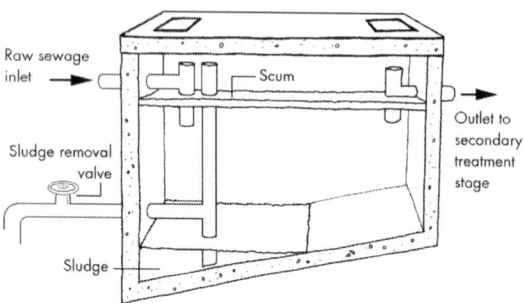

Figure 3.3. Cross-section of a settlement tank.

There is little to go wrong other than blockages, which are common to any primary stage. However, dip-pipes, crucial to preventing carry-over of excessive solids that would overload the leachfield or treatment system, can be damaged or knocked off by careless desludging. Very often system failure is blamed on the septic tank, although it is usually the leachfield that has become blocked or overloaded.

Pros:
- Little to go wrong.
- Established technology.
- Low head loss.
- Underground, so almost invisible.
- Prefabricated tanks can be installed in less than a day.
- Low cost compared with other forms of primary treatment.

Cons:
- Often misunderstood.
- Provide primary treatment only; must not discharge directly to watercourse.
- Must be desludged regularly (despite stories to the contrary).
- Effluent is anaerobic, so smells (this need not be a problem if the effluent is contained and proper venting is in place).

3. Solids separators (primary treatment)

This heading has been created to include a Swedish invention called the Aquatron (figure 3.4) and is best understood by referring to the diagram (figure 5.5 on page 85). Toilet waste enters the top of the hourglass shaped unit, losing momentum in the upper chamber before falling through the

Figure 3.4. The Aquatron solids separator. (Elemental Solutions).

neck. Solids fall straight down into the compost chamber directly below, whilst the liquid clings to the plastic neck wall and flows into the lower chamber of the separator and exits via the outlet pipe. When connected to toilets only, the liquid effluent is surprisingly unsullied and odourless. When connected to drains also containing greywater, the Aquatron effluent is typically equivalent to septic tank effluent in quality but free of unpleasant odour and septicity. Great care is required in the design and installation of the sewer leading to the Aquatron and the approach is best suited to single household and low population situations.

Pros:
- Provides primary treatment.
- Aerobic effluent (relatively odourless).
- Compost generated (no sludge).
- Can be installed above ground or in a cellar, basement or church crypt.

Cons:
- Requires a fall of 0.5m plus compost chamber – 2m in total.
- Composting requires ongoing awareness and maintenance (for example regular addition of soak material).
- Frequent cleaning of the Aquatron is essential to retain effective separation.
- Suitable only for the "sewage-enthusiast".
- Not a standard technology in the UK so may be unfamiliar to builders and regulators
- Fats remain in effluent.
- Potential for toxic shock to subsequent treatment stages, as no buffering involved (cf 12 hour average retention time in septic tank).

4. Percolating filters (secondary treatment)

'Filter' is a bit of a misnomer but these systems are also known as trickling filters, biological filters, biofilters, clinker beds, rotating arm systems, bacteria beds and filter beds. Percolating filters (figure 3.5) are usually preceded by a primary settlement stage and followed by a humus tank.

A percolating filter is a container, typically 1.2 to 2.0m deep, filled with a high surface area material, such as blast furnace clinker, stones or moulded plastic. This material is generically known as the media or bed matrix. Sewage is distributed over the surface of the media and drains freely from the base. Distribution is typically achieved by a rotating arm driven by the in-coming water. The sewage clings as a thin film to the surface of the media. In this thin nutritious layer, surrounded by plenty of air, microorganisms thrive and multiply, forming a slime known as the biofilm or zooglial film (after one of the commonest bacteria in the film, *Zoogloea*). It is the contact with the microbes in the biofilm that cleans the water, the bugs digesting the various contaminants in the water. As this biofilm gets thicker the inner layers of bugs attached to the stones are smothered and die and the biofilm is loosened from its anchor. This process of 'sloughing off' is aided by a range of larger organisms that graze on the film. The sloughed material can be removed from the liquid in a secondary settlement tank, leaving a clear liquid fit for discharge (as described in chapter two).

As with other biological treatment systems, the microorganisms can be killed if poisonous substances are put into the sewage system. In this respect trickling filters are generally considered more robust than activated sludge or package plants but less biologically robust than vertical flow reed beds.

Sewage Treatment Systems   29

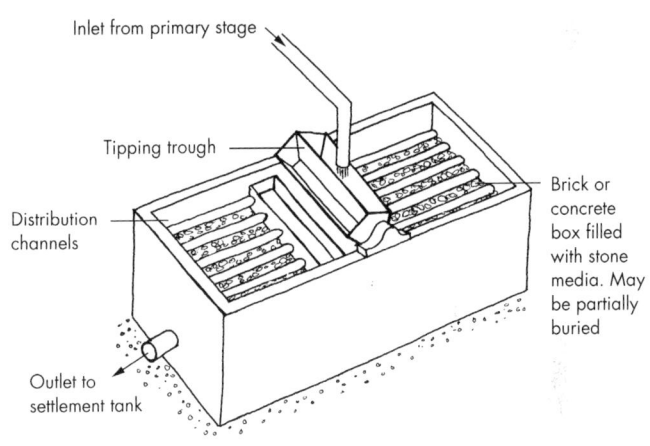

Figure 3.5.a. (top) Percolating filter and primary settlement tank for a small rural community.
Figure 3.5.b. (above) Schematic of a small domestic percolating filter with tipping trough.

Sizing is covered by BS.6297.
   Pros:
   • Relatively robust.
   • Tolerant of peak loadings.
   • Tried and tested.
   • Does not require power if a fall is available.
   • A 'green' technology despite the lack of plants!

Cons:
- If the distribution mechanism jams, treatment effectively stops.
- Expensive to install, unless self-built.
- Rarely used today for single household systems.
- Needs a fall of 1.5–2m or a pump.
- Regular de-sludging of humus tank is crucial.
- Influent is smelly, so these filters are usually required to be placed some distance from dwellings.
- Biofilm will die off if system is unused for long periods
- Visible (but need not be an eyesore).

5. Package plants (secondary, usually include primary)

Package plants (figure 3.6) are 'off-the-shelf' treatment systems for treating raw or primary treated sewage. Package plants can be installed quickly, often in a day, and are relatively compact. They are widely used and well established, with maintenance support usually available from the supplier. The main variations are outlined below. Almost all use an electrical power input but recent developments include systems that work by gravity where suitable fall is available

Features common to all package plants:

Pros:
- Compact.
- Fast installation.
- Readily available.
- Models with certification and approval available.
- Do not require much land or fall.
- Medium cost (comparatively) for secondary treatment.
- As close to 'fit and forget' as is presently available.
- Maintenance contract often available from installer.
- Can be unobtrusive as mostly buried in ground.

Cons:
- Most use electricity (amount varies considerably between models).
- Require regular maintenance, including costly replacement of parts for proper function.
- Slight noise and odour of some designs may cause annoyance.
- Not generally tolerant of fluctuating or intermittent loads (especially if sudden).
- Small size limits buffering effect.

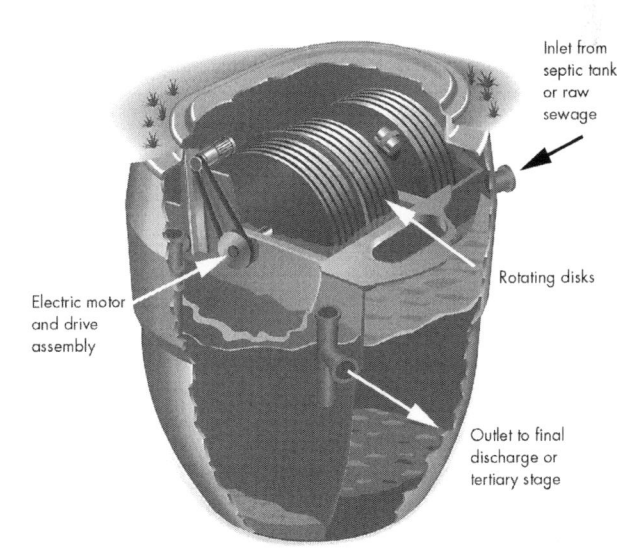

Figure 3.6.1.a. (top) Rotary biological contactor (installed below surface).
Figure 3.6.1.b. (above) Cutaway of a rotary biological contactor (installed below surface)
(courtesy of Klargester Environmental Engineering Ltd.).

- No treatment in the event of a power cut or mechanical failure (following which start-up can be difficult).
- Large amount of secondary solids may be created (design dependent).

Other Comments:
- May incorporate primary settlement in same package.

- Once-to twice-yearly sludge removal required.
- A reed bed can be used to improve effluent quality and overall system robustness
- Discharges effluent, usually to a watercourse.

5.1 Rotary biological contactors (RBC)

Also called RBC's or Biodiscs (figure 3.6.1), these feature a series of parallel, high surface area, plastic discs mounted on a horizontal revolving shaft driven slowly by a motor. A biofilm develops on the surface of the discs, which dip into the sewage. As they turn, the biofilm is exposed to air, providing oxygen for aerobic degradation of the sewage by the microorganisms.

Pros:
- Tried and tested.
- The better makes are reliable (no pumps, compressors or timers to fail, or distribution channels, pipes or nozzles to block).
- Quiet.
- Typically the lowest power consumption of the available package plants.
- With recirculation, very high effluent quality (including nutrient removal) possible.

Cons:
- Some mechanical parts are manufacturer specific.
- Relatively costly replacement parts.

5.2 Recirculating biological filters

These units are similar in concept to percolating filters except that they employ lightweight plastic media housed in a plastic casing. Effluent is recirculated over the media by electric pump so no fall is required and the unit can be made smaller than the equivalent percolating filter. Some designs use a conventional submersible pump and others use bubble lift pumps with an air compressor (figure 3.6.2) but the principle is the same.

Pro:
- The only mechanical parts are the pump and/or compressor, which are reliable and standard items.

Cons:
- Distribution system needs regular cleaning on some designs.
- Energy consumption can be high.

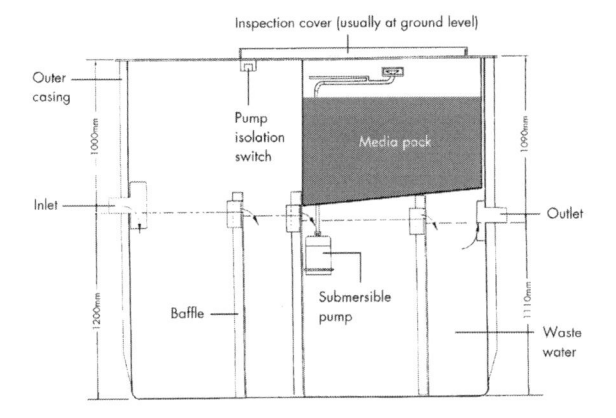

Figure 3.6.2 Cross-section of a recirculating biological filter *(courtesy of Burnham Environmental Services)*.

5.3 Submerged biological aerated fixed film system (BAFF)
This approach involves a process that is a hybrid of recirculating percolating filters and activated sludge (see 5.4 below). The units include a biological filter through which effluent passes and air is bubbled.
Con:
• Energy use typically higher than RBC.

5.4 Activated sludge package plants
These units make use of variations on processes commonly used in large scale municipal treatment works. The small package versions usually involve bubbling air through the incoming sewage (figure 3.6.4), oxygen being rapidly used to degrade contaminants. This process creates a slurry which contains microorganisms in the most rapid phase of growth and thus ideal for sewage breakdown. The slurry is allowed to settle, separating the active microbes as sludge from the comparatively clean effluent, which can be removed for further treatment or discharge. In traditional activated sludge systems, a proportion of the sludge is returned to seed the incoming sewage, hence 'activated sludge'. The remaining sludge accumulates and must eventually be removed.

There are several variations on this theme. Some packages involve two or three chambers for aeration and settlement, with sludge recirculation.

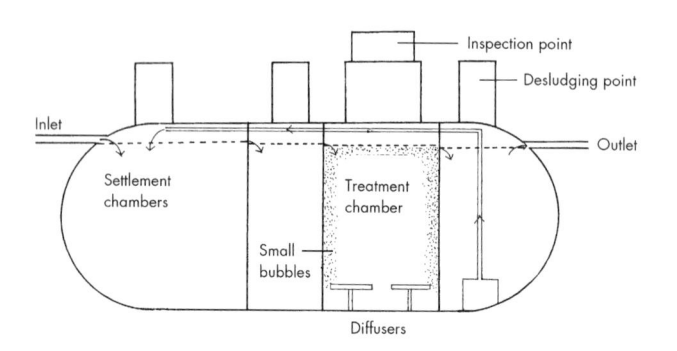

Figure 3.6.3. Cross-section of a submerged aerated biological filter (courtesy of Burnham Environmental Services).

Others — known as sequencing batch reactors (SBR) — conduct the whole process in a single chamber by temporarily switching off the aeration device to allow settling, then drawing off the liquid for discharge (fig. 3.6.4.b).
Pros:
- Some units aerobically digest the primary sludge.
- Very high quality effluents possible including nutrient removal.
- Some models can accept varying loads.

Cons:
- Compressors and air filters require maintenance.
- Under- or over-loading can lead to sudden 'dumping' of sludge blanket from plant to watercourse.
- Energy use is typically highest of all package plants.

Other comments:
- Performance varies between designs.

5.5 Fibrous media biofilters

This is a term we made up to cover a treatment plant that utilises a fibrous media such as waste peat fibre (a waste product from peat milling for power stations, an interesting ecological dilemma). Settled effluent is dosed over the surface by pump or gravity fed tipper or other dosing device. The fine fibres provide a lot of surface area and reduce the risk of drying out during periods

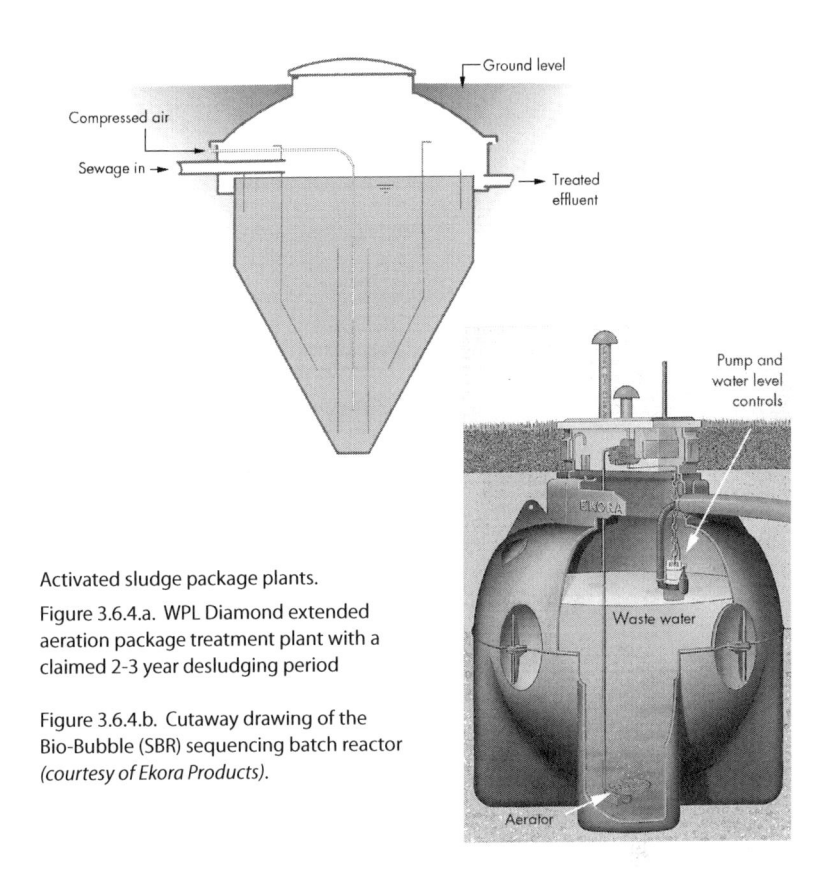

Activated sludge package plants.

Figure 3.6.4.a. WPL Diamond extended aeration package treatment plant with a claimed 2-3 year desludging period

Figure 3.6.4.b. Cutaway drawing of the Bio-Bubble (SBR) sequencing batch reactor *(courtesy of Ekora Products).*

of rest. Even where the settled sewage or resulting effluent must be pumped, the energy use is far lower than for other package treatment plants.

Pros:
- No secondary sludge produced.
- Very high quality effluents claimed including nutrient and pathogen removal.
- Tolerance of varying loads (schools, holiday homes, visitor centres).
- Only "mechanical" components are simple drainage and distribution pumps.
- Some are modular and so can be extended to cope with increased load.

Cons:
- Less compact than other package plants due to the more passive operation.

- Life of media finite.
- Some are above ground and so require visual screening.

Other comments
- More intensive versions exist but the authors remain to be convinced of claims of superior efficiency.

Figure 3.7. The Puraflo system serving a visitor centre. This is an example of a fibrous media biofilter. (Elemental Solutions).

6. Vertical flow reed beds (secondary & tertiary – occasionally primary)

Vertical (down-)flow reed beds (fig. 3.9) are usually preceded by some form of primary treatment, although some have been built to receive raw sewage (favoured in France). In terms of the direction of flow through these beds, they resemble percolating filters except that the bed media comprise a range of different grades of gravel, coarsest at the bottom, with a sand layer uppermost, planted with aquatic plants, usually common reed (Phragmites australis). The wastewater is spread out over the surface of the bed and then percolates down through the sand and gravel media and out at the base. Intermittent 'dosing' of the bed with the effluent by pump or gravity-operated flushing device (figure 3.8) improves surface distribution, as well as allowing periods of rest, which increase aerobic treatment by the bugs. Unlike the other reed beds, there is no depth of water standing in this type of

Figure 3.8. Gravity-operating flushing device.

reed bed. The percolating effluent encounters biofilm bathed in the air spaces between the matrix particles and roots, which accounts for the high degree of aerobic treatment. Deep sand layer beds (>500mm sand) eliminate the discharge of secondary solids, avoiding the need for subsequent treatment stages, and allowing direct discharge to watercourse.

As well as providing physical filtration the sand layer supports active biological treatment. The sand also improves surface effluent distribution and keeps the bed moist during long periods of rest, an advantage where use is intermittent.

Beds have been built using live willow structures and organic media such as tree bark or mature compost and this offers an even more low cost, ecologically acceptable solution in situations where complete containment is not crucial.

Pros:
- Established technology, included in latest Building Regulations.
- High levels of treatment possible, including nutrient removal with recirculation.
- DIY possible, which can be economical.
- Needs no power if a gradient is available.

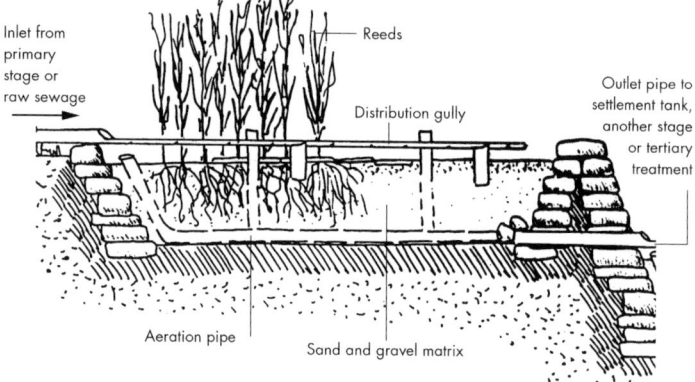

Figure 3.9.a. (top) Vertical flow reed bed (Elemental Solutions).
Figure 3.9.b. (above) Cross-section of a vertical flow reed bed.

- Needs no liner if land is clay-rich, and water table low.
- Can be very attractive to the eye.
- Generates interest, enthusiasm and awareness of sewage.
- Negligible secondary solids discharge.
- Maintenance is technically simple.
- Biologically complex and robust.
- Performance maintained, even with sudden and large load fluctuations.
- Failure tends to be gradual, allowing time for preventative action.
- Works well even before plants are established; better when mature.

Cons:
- Requires specialist design.

- Requires a fall of at least 1.5m to provide good treatment.
- Requires more space than conventional systems (notably package plant).
- High water table situation requires pumping and/or bed lining.
- Not cheap unless DIY.
- Often seen, wrongly, as a 'green' panacea.
- Sensitive to hydraulic overloading.

Other comments:
- Seek advice before DIY
- Sand grading and bed sizing are critical to avoid blockage.
- First generation required two stages of parallel beds; second generation single bed approach (for example compact vertical flow reed bed) increasingly used since 1997
- Requires awareness similar to gardening.
- Requires approximately 2m^2 per person served.

7. Subsurface horizontal flow reed beds
(tertiary but sometimes secondary)

The subsurface horizontal flow reed bed (figure 3.10) is characterised by sewage flowing horizontally through the gravel (occasionally soil), as well as the plant roots and rhizomes. The arrangement can be likened to a bath, filled with gravel and planted with aquatic plants. Once you have filled the bath, water overflows at the far end. Thus, a depth of water of some 30-50 cm is maintained in the bed, unlike in vertical flow beds which are free-draining (bath plug removed!). Since oxygen finds it difficult to diffuse through the depth of water, less oxygen is available for aerobic treatment. Hence, there is poor BOD removal from strong effluent and almost no ammonia removal.

Whilst such reed beds are occasionally used for secondary treatment of sewage, the presence of high levels of organic matter, the low levels of oxygen, and the tendency to block (requiring replacement of the bed media and replanting) make horizontal flow reed beds better suited to tertiary treatment. In fact, we strongly recommend *not* attempting to treat septic tank effluent with this type of reed bed (contrary to what is allowed by UK Building Regulations). That said they can do an excellent job removing fine particles of organic matter that are too small to be removed in a settlement tank. Adequate treatment before the horizontal flow bed will extend the bed's life to 8-10 years; but we must emphasise that gravel replacement and replanting are almost certain to be required eventually.

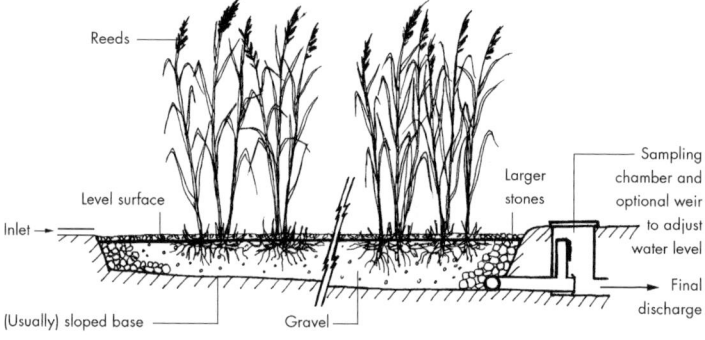

Figure 3.10.a. (top) A horizontal flow reed bed at Newport, Shropshire (4,000 pop.).
Figure 3.10.b. (above) Cross-section of a horizontal flow reed bed.

Pros:
 • Hundreds of horizontal flow reed beds are now in operation in the UK.
 • Becoming established technology, included in latest Building Regulations.
 • Often low cost.
 • Natural looking.
 • Need less fall than vertical flow beds.
 • Provide buffer before the discharge.
 • Minimal maintenance during operating life.
 • Can use a wide range of water plants.

Box 3.1. So what do the plants do?

Whilst there is still some difference in opinion we offer the following suggestion as to the function of reeds in reed beds. (Where it is specific to one or the other we indicate VF for vertical flow and HF for horizontal flow.)

Aquatic plants DO:

- Provide beauty, making people more likely to care for the system.
- Blow in the wind, opening up the sand surface of the bed (VF).
- Insulate the bed, providing wind and frost protection.
- Transport air to their roots, which may enhance treatment locally.
- Evapotranspire, so helping (VF) bed regeneration when resting.
- Reduce growth of weeds, i.e. plants that may be less beneficial than reeds.
- Encourage worm activity (VF), increasing surface permeability.
- Increase the C:N ratio of surface film, so accelerating composting (VF).
- Stabilise bed surface (VF).
- Support larger densities of beneficial bacteria than does the surrounding media, which may be significant.
- Release selective biocidal compounds, killing harmful microbes, although the significance of this may be small.
- Provide physical filtration and electrostatic attraction of small particles (HF).
- Provide a habitat for microorganisms and larger consumer organisms.
- Take up some heavy metals (HF).

Aquatic plants DO NOT:

- Provide net oxygenation of wastewater (unless already well treated).
- Increase hydraulic flow, due to dead and decaying rhizomes.
- Take up a significant proportion of the dissolved nutrients from the wastewater (in the cool UK climate).
- 'Eat sewage'.

- Robust.

Cons:

- Often seen wrongly as a green panacea.
- Often misunderstood to be capable of treating strong effluent and ammonia.
- Prone to blockage and odour generation.
- Requires gravel replacement and replanting, eventually

Other comments:

- First UK beds installed in 1985.
- Recommended as a tertiary treatment stage.

- Contrary to many official guidelines the authors consider that this type of reed bed is *not* suitable for *secondary* treatment.
- The role of reeds has been overstated.

8. Free water surface flow reed beds

Also known as overland flow reed beds or surface flow reed beds. As the name suggests these reed beds, unlike the previous two, have plants emerging from a depth of water, with no gravel media. They use a simple excavation, filled with water to a depth controlled between 100-300mm, which overflows at the far end, allowing settlement of solids, as well as their entrapment in the plant stems and accumulated biomass. Oxygen diffuses through the open water surface allowing aerobic metabolism, including nitrification. In addition, dentrification occurs in the anoxic mud in which the plants are rooted. Although only a handful have so far been installed in the UK for sewage treatment, they are much the most common type of constructed wetland treatment in the USA (where Phragmites australis is almost never deployed!), where land area is more abundant and where engineered natural wetlands are also used for treatment in certain cases.

Pros:
- Achieve similar tertiary treatment to subsurface horizontal flow beds but without gravel media.
- Excellent wildlife habitat for nesting and (unlike other bed types) for feeding, akin to natural wetlands.
- Extremely simple and cost-effective construction.
- Easy maintenance, as no gravel to become blocked.

Cons:
- Require about twice the area of subsurface horizontal flow reed beds for same treatment.

Other comments:
- Little used in UK to date.
- Authoritative design information available from USA installations.

9. Ponds (primary, secondary, tertiary)

These are also known as waste stabilisation ponds, settlement ponds, lagoons, or sewage ponds. Original installations were in hot countries with a small anaerobic pond at the start, followed by larger aerobic ponds. Because of this tropical origin, any plants that colonised the pond edges were removed, as their stems were insect breeding sites. Pond systems are now well

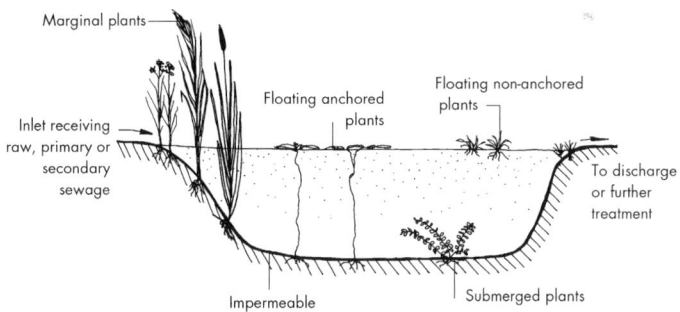

Figure 3.11.a. (top) Planted pond for village of 150 people in Shropshire (Elemental Solutions).
Figure 3.11.b. (above) Cross-section of a solar pond.

established as having a role in cooler climes, with thousands of such systems long established in the USA and northern Europe. Most of these serve large populations and are vast, somewhat brutal, concrete constructions. In the UK, recent practice has seen many ponds deliberately seeded with a wide range of aquatic plants (figure 3.11) often in tandem with reed beds, making the systems places of great beauty. These are often simply termed planted ponds.

Like free water surface flow reed beds, oxygen for the treatment organisms is provided by diffusion from the air over the large surface. In addition, with plenty of light reaching the main body of water, photosynthesising algae living in the pond produce a lot of oxygen. Although a large surface area is required to ensure sufficient treatment, smaller versions of these systems have been made possible by pumping the water to recirculate via aeration cascades or by using wind powered mixing devices. Due to their great volume, ponds can absorb shock loading well. Unless there are leaks, ponds rarely fail.

Pros:
- Robust.
- Can provide primary, secondary and tertiary treatment.
- Resistant to temporary organic and hydraulic overload.
- Can be beautiful.
- Pathogen removal can be excellent.
- Sludge removal is very infrequent.
- May not need liner if clay is present.
- DIY possible.
- Wind-powered aeration devices possible.
- Excellent wildlife habitat.

Cons:
- Require large area (approx 10–20m^2/person).
- Occasional odours can be a problem if first stages are not aerated.
- Requires power if aerated.
- Energy consumption, if aerated, is typically far higher than for an equivalent package plant.
- Effluent may contain high levels of algae.
- Sludge removal may be difficult.
- High installation costs for smaller systems.
- Needs special health and safety considerations.
- Plastic liners can be vulnerable.

10. Leachfields and soakaways (secondary, tertiary and dispersal – occasionally primary too!)

A leachfield is used as the last part of a treatment system. It is usually preceded by a septic tank and this combination is often referred to as a 'septic tank system'. A leachfield is a series of perforated pipes, surrounded by gravel, that run in underground trenches (figure 3.12). With a well designed leachfield in suitable soil, the wastewater is thoroughly cleaned within about a metre of travel through the soil.

In areas with very porous soils, a simple pit full of rocks is often used to receive the wastewater. These 'soakaways' or 'leach pits' do not ensure passage of water through a sufficient area of soil and so treatment is much

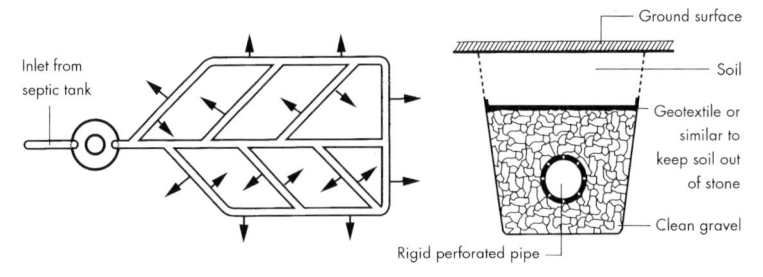

Figure 3.12.a. (top) Leachfield (installed in the foreground) (C.Weedon).
Figure 3.12.b. (above left) Plan view of herringbone leachfield used where single pipe would be too long.
Figure 3.12.c. (above right) Cross-section of a leachfield.

reduced. It can be argued that such soakaways should be avoided because the untreated wastewater may find its way into the water table, or even a drinking water supply.

Failure of the leachfield usually shows as a wet smelly patch on the lawn or through tank overflow (check for inlet blockage first). This may be caused by long term accumulation of solids or 'sewage sickness', in which case often the best solution is to dig a new half-size leachfield away from the old one. This can be used whilst the original field rests and regains its permeability. The two fields can then be alternated on a yearly basis. If the blockage was due to the discharge of gross solids, for example caused by overdue sludge removal or a broken dip pipe (see septic tanks, figure 3.2 above), then it would be worth having the leachfield pressure-jetted by a contractor once the cause has been cured. If it works this is the cheapest solution.

Some innovative leachfields have been created using a wide half-pipe (450mm) for distribution and dispersal. These 'Trench Arches' have been used without primary settlement preceding them for intermittent discharges such as from rural churches – see the Case Studies for an example.

Pros:
- First choice for on-site sewage disposal system, in terms of both cost and low environmental impact.
- Very well established.
- Provides treatment and disposal.
- May require a large area but this can be under a lawn or similar.
- No odour when properly specified and installed.
- Invisible.
- Excellent treatment when correctly specified and installed in appropriate situation.
- Little or no maintenance.
- Lowest cost secondary treatment.
- Natural treatment in its simplest and most efficient form.

Cons:
- Fissures in the ground or land drains can allow untreated effluent to reach a watercourse or well.
- Ground must be suitable and the water table must remain at least 1m below the trench bottom.

Other Comments:
- An old existing leachfield may not cope with increased water usage.
- Often unfairly criticised by salesmen with vested interests.

11. Willows and trenches (mainly tertiary treatment and dispersal)

When sewage has been treated to a sufficiently high standard one may wish to discharge the effluent to the surface of the soil as an irrigant. For the sewage this is a very useful final 'polish', since the topsoil is even more efficient than subsoil at 'absorbing' nutrients and the final suspended matter. For the soil, it is a source of both moisture and nutrients, and allows the growth of a crop. The authors' experience is only of growing trees — mainly willows, growing between shallow irrigation channels (figure 3.13; see chapter seven for details) – but most plants will thrive. Willows are ideal, not only thriving in the wet, nutritious conditions but yielding a useful harvest, suitable for

Sewage Treatment Systems   47

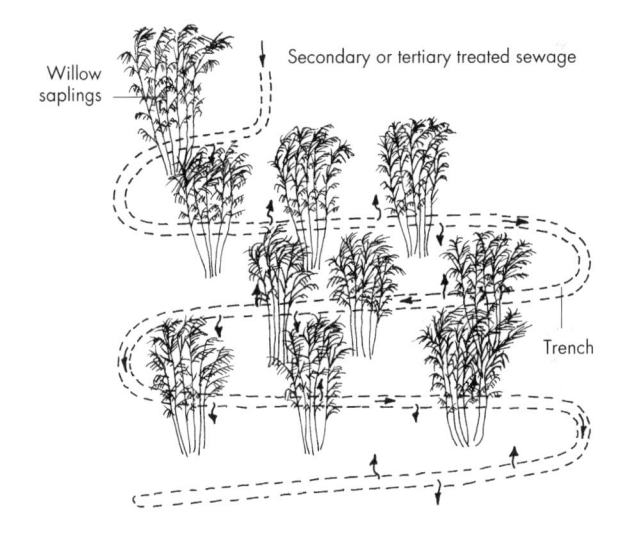

Figure 3.13.a. (top) Willow and trench treatment stage (with sheet mulch) (Elemental Solutions).
Figure 3.13.b. (above) Schematic of a willow and trench treatment stage.

fencing, bean poles, fuel, basketry and so on, or even for planting for further growth.

Pros:
- Provides tertiary treatment and often disposal to ground.
- Produces a useful crop.
- Can be beautiful.
- Good wildlife habitat.
- Ideal DIY.
- Good after ponds for dealing with algae.

Cons:
- Can be a dull monoculture.
- Requires a large area of suitable land.
- Possible environmental health considerations.

Other Comments:
- Needs weed- and possibly pest-control if harvest is for craft use.
- Yield increased by nutrients but may be poor quality for basketry.
- Usually discharge standard must be met before this stage.
- Will not dry up wet ground in winter.
- Discharge from the willow bed is likely for parts of the year, in impermeable soil.

12. Living machines (primary, secondary, tertiary)

This name, coined by Ocean Arks International (an organisation established by Dr John Todd, the main developer of this approach), refers to artificial ecosystems, constructed in a carefully formalised manner. Their design is based on a number of principles intended to maximise the similarity of the constructed ecosystem to its counterpart in nature. A key feature is the seeding of the system with selected organisms, minerals and other materials, the aim being to lead to a self-designing dynamic equilibrium. Among other assertions the enormous surface area of the plant roots involved is claimed to be essential and to provide effective cleaning. However, a relatively large power consumption and high human input into operating the systems actually installed seems to counter the claimed ecological benefits.

Most early Living Machines (figure 3.14) were pilot municipal treatment systems in the USA. A few have been installed in the UK to treat domestic sewage on a small community scale, with mixed results!.

Figure 3.14. A 'Living Machine' in a greenhouse, Providence, Rhode Island, USA (Elemental Solutions).

Pros:
- Visually interesting.
- High levels of treatment claimed.

Cons:
- Extremely capital-intensive in relation to other systems.
- High energy input (aeration, pumping and sometimes frost protection).
- Low efficiency.
- Many implicit and explicit claims have been found to be unsubstantiated.
- Require close management.

Other Comments:
- A concept and design philosophy rather than a specific technology.
- Has gained a 'green' image but actual performance credentials of potential installations should be carefully explored.

13. Living soakaways

At sites with plenty of gently sloping land but very shallow topsoil lying above clay or rock a conventional leachfield will fail. One unconventional approach that has been shown to be successful is this situation has been named a living soakaway, or planted soil filter. Typically an area of land is cultivated with the addition of large quantities of soil-improver such as composted wood chip. Effluent from a septic tank is pumped or dosed

50 Choosing ecological sewage treatment

Figure 3.15. Example of a newly planted 'living soakaway' serving a rural office complex. (Elemental Solutions).

into an irrigation pipe running along the contour at the highest point. The irrigation pipe is suspended in a half pipe and the whole area is covered in a deep mulch of wood chip or bark to absorb any odour. As there is no stone used it is possible to make a very large infiltration area for little cost provided land is available. The area can be planted with coppice trees as there are no pipes in the ground to be blocked by roots.

Pros:
- Very cost effective where suitable land is available and coppice is desired anyway.
- Excellent solution for heavier soils where there is no watercourse to discharge treated effluent.

Cons:
- May not be familiar to planners or environmental regulators.
- Not covered by Regulations or Standards and so must be designed by a specialist who can take professional responsibility for it working.

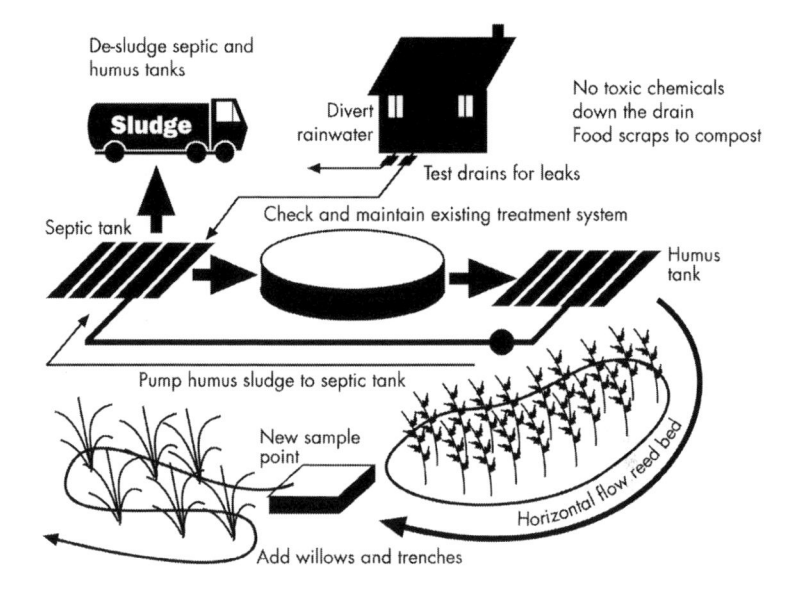

Figure 3.16. Upgrading your existing sewage treatment system.

If it ain't broke don't fix it – upgrading your existing sewage treatment system

It is not unheard of for people, perhaps bitten by the reed bed bug, to rip out intact but underachieving sewage treatment systems and replace them with a sexy new reed bed. This is, after all, how two of the authors got into sewage! In hindsight, knowing a little more than we did then, we can see that we could have saved a lot of money, time and hassle by fixing what we already had rather than ripping it out.

The offending sewage system was a percolating filter that was failing its consent on BOD and suspended solids. The first mistake we made was not to realise that the samples the inspector was taking were not wholly sewage; the outlet pipe was broken, so it acted as a land drain collecting mud and dissolved cow pat from the field it passed through. Since this was not discovered until we had served our sewage apprenticeship, we will never know how badly the original plant was actually working but it is very likely that it could have been brought up to standard relatively simply (figure 3.16).

The two main weaknesses of the old filter were that it received rainwater from a roof and courtyard and that the humus tank was undersized. The

> **Box 3.2. Before upgrading — checklist**
>
> • Check roof and other rain water does not enter system.
> • Check drains for leaks after heavy rain.
> • Check for dripping taps and other water wastage that might overload the system.
> • Check septic/settlement tanks for broken dip pipes and sludge level.
> • Check biological treatment plant for correct function and service as recommended by supplier.

first of these had to be rectified before the reed bed was built anyway, and the second could have been fixed by the addition of an extra tank. We would then have had a reasonable system that, probably, would easily have met the required legal standards.

The moral of this cautionary tale is not to rush into ripping out an existing system even if it is failing. First evaluate the problem (see box 3.2), aided by your new-found understanding and enthusiasm for sewage treatment. Then weigh up the pros and cons of the various solutions and make an informed decision. You may find a suitable fix for your existing system, or you may discover that your treatment plant is near the end of its useful life, uses large amounts of electricity, and that another of the systems described above is just the thing.

Whilst detailed design and troubleshooting are beyond the scope of this introductory book, it should give you the tools to assess whether your existing treatment plant is likely to be worth saving. Also you should be able to spot if someone is trying to sell you something you don't need.

Summary

On-site treatment of sewage can be achieved by combining any of a wide variety of methods. Both conventional and 'alternative' methods have appropriate application, no single approach being suitable in all situations.

The upgrading of an existing system that is apparently under-performing may well be best served by a combination of modified water use, repair of the system and/or the addition of a single treatment element, rather than its replacement with a whole new treatment system.

As a client or specifier you should be wary of manufacturers' unsubstantiated 'green' claims.

Chapter Four
Monitoring and Regulations

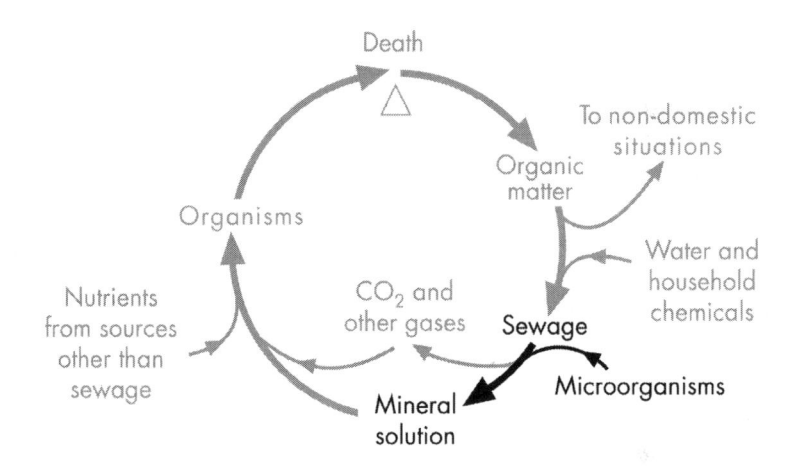

So far we have discussed the context, the processes and the hardware for on-site sewage treatment. But, whether you choose to revive an existing system or to install a new one, or if you simply want to assess how an existing system is performing, some method of monitoring treatment success is necessary.

This chapter deals with some of the monitoring methods and attempts to explain what the results of such approaches really mean:

1. Monitoring methods
2. Regulators and regulations.

Monitoring methods
We can identify four approaches to water quality assessment, of increasing sophistication:

1. The jam-jar method.
2. The turbidity tube.
3. Biochemical tests.
4. Biotic indices.

The jam-jar method

A simple way of checking whether your system is functioning is to collect a jam-jar of the effluent and observe and smell the contents (see figure 2.2 on page 19). From this you will get a fairly good, instant feedback that will become more and more reliable as you gain experience. The colour will let you know whether there are algae present, and grit and pebbles will be readily apparent. The cloudiness is an indicator of both suspended solids and BOD. A clear liquor with solids that settle after a little time indicates that the system is working biologically but that the secondary settlement is inadequate or in need of de-sludging. However, the jam-jar technique is subjective and not suitable for gathering data you can compare from day to day or as a defence against prosecution by the regulators.

The turbidity tube

A simple step towards quantifying such monitoring is the turbidity tube (figure 4.1) — a long thin transparent tube with calibrations down its length and a black cross on the bottom of the tube. Look down the cylinder to the cross and pour in treated sewage until you can no longer see the cross. Then read off the height of the liquid. That is the turbidity reading. A long transparent column of effluent, filling the tube, indicates clean water; a shallow depth (before obscuring the cross) indicates dirty water. The figures obtained from one reading can be compared with those from others. But the method is still subjective; different people, and/or light conditions yield only approximately correlating results.

Figure. 4.1. The turbidity tube

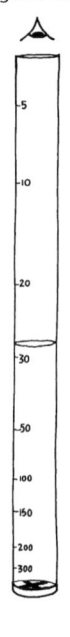

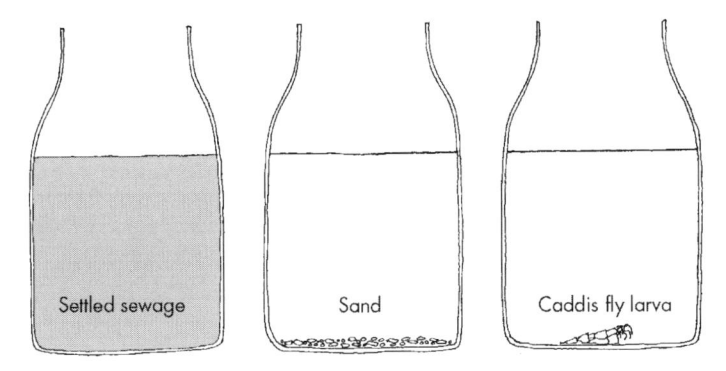

Figure 4.2. Water-borne solids of differing character but giving the same suspended solids measurement.

Biochemical tests

The methods described above are limited by their lack of reproducibility and the difficulty of valid comparison between different tests at different times and places. When setting standards that the public are asked to achieve by law, the regulatory authorities are obliged to utilise methods that do not suffer from these limitations. Scientific rigour is required. For this, specific analytical tests (usually, but not always, needing specialist training and expensive equipment) have been devised, each measuring a different characteristic of the wastewater. In order to fully understand sewage treatment, such that you are in a position to design your own system or make the best of upgrading an existing system, you would do well to arm yourself with a basic knowledge of these 'determinands'.

This is a somewhat technical section but nonetheless important when it comes to talking the same language as the regulatory authorities. It is arranged as follows:

1. The major determinands — SS and BOD.
2. Other determinands — nitrogen, phosphorus, sulphur, pathogens.
3. Interpreting biochemical analysis data.

The major determinands — SS and BOD

The regulatory authorities use two main determinands to set discharge standards to be achieved by small sewage treatment systems. Many other

Box 4.1. Primary and secondary solids

Sewage that has had nothing done to it is called unsettled or raw sewage. It contains a certain amount of material that could be intercepted in a laboratory filter and, therefore, gives a suspended solids reading. In the process of being treated, these lumps — 'primary solids' — are settled out of the flow of liquid or transformed and dissolved in the liquid. Thus the original SS gradually diminishes. However, the proliferation of mortal microorganisms in sewage inevitably means lots of micro-carcasses. These are susceptible to filtration and so can also register as suspended solids. These are generated in the sewage treatment process (rather than being present in the raw influent) and are distinguished from the primary solids by being named 'secondary solids'. UK law requires that the consent for suspended solids is achieved regardless of whether the solids are primary or secondary.

---

determinands exist that are considered to be of less importance (at the small scale) with respect to the health of natural water bodies. (See Other determinands, below).

Suspended solids (SS)

Let us start with the simplest to understand — suspended solids or SS (pronounced 'ess - ess'). It is a quick and cheap measure of the amount of solid material in the water.

If you were to fill a jar with a litre of water from your sewage outfall and then take it to a laboratory, the laboratory technicians would pass the sample through a filter of standard pore size and known weight. They would carefully heat the wet filter to drive off the moisture, and then re-weigh the filter. The difference between this measured weight and the weight of the clean filter is the weight of the suspended solids in one litre of effluent. SS is usually cited in milligrams per litre (mg/l), which is the same as parts per million (ppm) and g/m^3.

Roughly 60% of the organic load of domestic sewage is in the form of suspended solids. However, SS alone is not sufficient to indicate that there is a pollution problem, since it includes grit and other biochemically inert particles (see figure 4.2). Therefore, other tests are needed that, when viewed in combination with SS, can reveal more about the litre of water.

Biochemical Oxygen Demand (BOD)

The oxygen that microorganisms require in order to break down organic matter is called the biochemical oxygen demand. This mouthful is commonly referred to as the BOD (pronounced 'bee-oh-dee').

BOD is measured in milligrams of oxygen removed per litre of water over a certain period of time. The time element is important. For example, consider a tree. A tree has a huge BOD. The amount of oxygen required to digest an old oak tree that has fallen in a stream is enormous. But it will take years for all that digestion to take place and the demand for oxygen by the microorganisms will be spread out over those years. In any given period the oxygen removed from the water will be negligible compared to what is available in the flowing water.

Contrast this to rapidly decaying organic matter such as a cow-pat. The microorganisms would get to work very rapidly but they would be finished very quickly. During that short period of time, the oxygen in the water would be consumed faster than it can be replenished from the air.

In terms of depleting the stream of oxygen, the cow-pat is a greater challenge than the tree, even though the total oxygen requirement is less. Therefore, the figure of BOD set by the regulatory authorities usually has a little number after it – usually 5; hence BOD_5. This means that the amount of oxygen required by the degrading organisms is measured over five days in the laboratory incubator and is, therefore, strictly speaking a *rate* of oxygen utilisation, rather than a concentration.

Even worse than a cow-pat in a stream is a sewage discharge that continues day and night throughout the year. If salmon returning to spawning grounds encounter a cow-pat, they can loiter some distance downstream until the trouble is digested and dispersed. If they encounter the continuous discharge of a sewage outfall or a farmyard, the salmon will have to wait until the house becomes derelict or the subsidies are removed from milk production. Experience shows that salmon do not wait that long.

BOD is also dependent on temperature. The organisms degrading either the tree or cow-pat are much more enthusiastic about their task when the water is warmer. Therefore, the BOD is measured at a standard temperature (20°C) to make comparisons meaningful.

Finally, you may see BOD figures on technical forms cited as BOD_5 + ATU. ATU (allylthiourea) is a chemical added to inhibit another pathway of degradation that also requires oxygen ('nitrification' — see the discussion below).

To simplify further — and this is all you really need to grasp — big BOD is bad, little BOD is good. The same is true for SS. If you return the water at lower numbers than the levels set by the authorities, you will not be asked to explain yourself, improve your system or pay a fine.

Why is oxygen demanded?
It can be very helpful in refining elements of the design of a sewage system, and in becoming a creative designer for treatment of all kinds of effluents, to know where the demand for oxygen, or BOD, is coming from. The majority of microbial activity that contributes to this demand for oxygen is the metabolism of organic matter, which by definition contains carbon. The carbon itself binds to several other chemical elements as well as to itself, and it is the breaking of these bonds that requires oxygen.

An aide memoire for the six major chemical elements that make up living tissues is NCHOPS, the chemical symbols for (respectively): nitrogen, carbon, hydrogen, oxygen, phosphorus and sulphur. Combinations of these elements comprise the majority of living matter, along with smaller doses of other elements such as potassium and sodium, and even smaller amounts of trace elements such as zinc and cobalt. Space and degree of relevance force us to limit this discussion of biochemistry to NCHOPS.

In organic matter, these elements are found in various combinations to give fats (mainly carbon and hydrogen), carbohydrates (mainly carbon, hydrogen and oxygen) and proteins (mainly carbon, hydrogen, oxygen and nitrogen), plus many other types of molecule. Phosphorus and sulphur are found in lesser amounts in all three categories. Organic molecules are held together and dominated by long chains of carbon atoms. Biochemists refer to the carbon 'backbone' or the carbon 'skeleton' of a molecule. When organisms break down these molecules, certain bonds between atoms are broken, allowing energy to become available, which the microorganisms can utilise. Because oxygen can be used to allow this 'creation' of energy, a demand for oxygen — the BOD — exists.

As the bonds to the carbon atoms are severed, and the descending arc of the Big Circle is travelled, the incredibly long molecules that make up cabbages and other organic matter are degraded to smaller organic fragments, which have less energy stored within them. If oxygen is available, it is consumed in the process, combining with hydrogen atoms to form water. The fragments, too, will degrade until they become classified as mineral or inorganic.

Each of the six main elements in organic molecules will eventually arrive at the bottom of the circle in its own way. Carbon often emerges as carbon dioxide and is released to the environment as gas or remains dissolved in the water as carbonic acid. Hydrogen and oxygen are lost for choice as to which elements they will escape with (see box 2.2 on page 16).

Other determinands – nitrogen (N), phosphorus (P), sulphur (S), pathogens

Nitrogen and ammonia The sewage industry takes a very great interest in nitrogen and there is a lot of information about water quality to be gleaned from the measurement of nitrogen in its various forms. As we mentioned, nitrogen is to be found mainly, though not exclusively, in the proteinaceous material of tissues. As this protein is degraded, the nitrogen will follow a well-documented pathway. This starts within the body: urea is the main form in which N is excreted from our bodies and which degrades to ammonia.

Ammonia is simple to measure and is a frequently used determinand partly for that reason. It is also directly toxic to some creatures (young fish especially) and it is important to have a gauge of its concentration for that reason alone. However, ammonia analysis also lets the sewage worker know two things.

The first is whether carbon breakdown has proceeded to relative completion. It has been found that in most types of sewage treatment system, the microorganisms that transform ammonia 'wait' until the BOD has fallen to a large extent, before they start their work. In general we can say that if the ammonia concentration has dropped between the inlet and outlet of a stage of a treatment system, then substantial breakdown of organic matter has also occurred (and that can be ascertained even without taking a measurement of BOD).

The second thing that the ammonia reading will indicate is whether there is any oxygen available. The ammonia-transforming bacteria also require oxygen before they start their work. If there is no oxygen available, nitrogen will be preserved in the form of ammonia, whatever the BOD concentration.

Box 4.2. Five days

When the decisions about standardised tests were being made it was agreed to make the BOD incubation last for 5 days since this was calculated as about the average time for sewage to reach the mouth of the average British river. It also gave time for fairly complete degradation of the carbonaceous matter without too much oxygen used for nitrogen degradation in domestic sewage. Any shorter and the degradation would have a long way to go and any longer would not reveal much more about the samples. In a few countries this time is increased to seven days (fitting in better with the working week) and the paperwork shows BOD_7, rather than BOD_5.

Box 4.3. Other tests related to BOD

There are several other tests that are related to but significantly different from BOD. The most often seen is COD or chemical oxygen demand. This is useful when the test sample contains compounds that are not degradable by microorganisms, or not biologically degraded within the five-day incubation period. COD usually gives a higher figure than BOD because it includes all the material oxidised in the BOD test plus a host of other compounds. TOC (total organic carbon) gives an indication of the ability of the liquid to burn — an indication of the total organic and other combustible matter present. Like COD, this gives no indication of the source or rate at which the demand is required, so does not distinguish between the tree, the cow-pat, or even a fuel spill. The NRA (National Rivers Authority – replaced by the Environment Agency) was keen to introduce this determinand since it is quick and cheap and the hope was that if BOD and TOC were established for a given discharge the ratio between the two could be calculated. In future the simpler TOC test would then be a sufficient indication of the BOD.

Assuming that there is oxygen available and the carbon bonds are mainly broken, what would happen to the ammonia? If in high concentration, especially if the pH is significantly above neutral, ammonia will come out of solution and be lost as a gas. However, this is not the main route of ammonia loss from domestic sewage. Much more common is that ammonia is oxidised and the nitrogen becomes bound as nitrate, after passing swiftly through an intermediate stage of nitrite. This process is known as 'nitrification'. It is the process that ATU (see BOD, p.56) inhibits, since ATU blocks the action of the nitrifying bacteria. Nitrosomonas (and allies) change ammonia to nitrite and Nitrobacter (et al) take it on from there to create nitrate. This two-step process requires oxygen to be present (about 5mg for each mg of ammonia that is transformed) and is, therefore, a contributor to BOD unless ATU or another nitrification inhibitor is present. Nitrification also requires some inorganic carbon source for the bacteria's cell synthesis. This source is usually carbonic acid (at the rate of about 9mg for each mg of ammonia transformed).

Nitrifying bacteria also require some heat and do practically nothing in winter temperatures (<5°C), which is why most sewage systems are susceptible to returning high ammonia figures in the cold. So you can see that for ammonia transformation there are many conditions that need to be right: temperature, low BOD, the presence of oxygen and inorganic carbon sources, and not too much nitrate, ammonia, or ATU.

Assuming, however, that this set of circumstances is met, nitrate is the

Box 4.4. Milk

Would you prefer a discharge of milk or of hydrochloric acid into your favourite trout stream? If one were simply to consider BOD one would in fact prefer the acid since bacteria do not thrive in it and thus they would demand no oxygen. Milk, on the other hand, is transformed rapidly and profoundly by microorganisms; one only needs to leave a bottle on the window sill for a few warm hours to see that this is so. The microorganisms would consume about 150,000 mg (0.15 kg) of oxygen for every litre of milk they sour, over 5 days at 20°C. Hydrochloric acid is harmful in other ways, creating conditions too acidic for normal biological activity. This shows the importance of analysing a range of parameters — i.e. the 'suite' of determinands — for a sample, if its full pollution potential is to be revealed.

Examples of average BOD_5 in mg/l:

Tertiary treated sewage	2 — 20
Secondary treated sewage	10 — 20
Raw sewage	100 — 400
Farmyard washings	1,000 — 2,000
Vegetable washings	3,000
Cattle slurry	10,000 — 20,000
Silage effluent	12,000 — 80,000
Milk	150,000

---

form in which virtually all nitrogen from the original organic compounds is found in the mineral solution at the bottom of the Big Circle. That is to say that if a sewage works has done this part of the job well it will discharge nitrates. There are of course many problems associated with nitrate production and disposal but poor treatment by the sewage plant is not usually one of them. So, we can summarise that urea discharge is unlikely (because it breaks down spontaneously); ammonia discharge is undesirable; and nitrate discharge indicates good BOD & ammonia removal (conversion), meaning that the wider water environment does not have to use up precious oxygen to break down our sewage.

So, what fate awaits nitrate after discharge? You will recall that we called the process of conversion of ammonia to nitrate 'nitrification'. The process of degrading or removing nitrate is called 'denitrification' and cannot take place until nitrification has taken place. The chemical formula for a nitrate ion is NO_3^-, indicating that each nitrogen atom is bound with three oxygen atoms. That oxygen can seem mighty attractive to certain bacteria if there is no other free oxygen around. (Conditions where there is no unbound oxygen are known as 'anoxic'.) A gagging microorganism can get to thinking that that nitrogen is too damn greedy for its own good in times of oxygen hardship and better start sharing. In these anoxic conditions, denitrifying bacteria are able to get the oxygen they need from nitrate leaving NO_2^-, (nitrite), NO (nitrous oxide) and N2 (nitrogen gas). The latter is by far the most abundant product of denitrification, the others being unstable intermediates. N_2 gas thus leaves the water and joins the free nitrogen that makes up 79% of the air we breathe. It is a totally harmless product of nitrogen metabolism.

A second means of removing the nitrate from solution is by having it taken up into the roots of plants as part of their living tissue or 'biomass'. However, even in reed beds, in our climate, this is a minority route for removal of nitrate from sewage wastewater as compared to what can be achieved by making unbound oxygen scarce. (In warmer climates there is evidence of significant contributions to the removal of nitrate by the uptake into regularly harvested plants.)

Therefore, to assist nitrogen compounds efficiently through their path towards mineral soup you must ensure that your sewage treatment plant first allows plenty of air into the water, in order to allow nitrification. After that you have a choice: you can take the water through a low oxygen environment, to allow denitrification (that is, provided there is some carbon present – didn't say it wasn't complicated!); or you can take the water through an aerobic environment in which plants can remove some of the nitrate by root uptake.

Biochemical analyses are used to measure the various nitrogen-containing molecules, either singly or in groups. Note that analytical results usually record the amount of a single element (for example N) in a molecule (for example NH_3, NO_3^-), in order that the amount of that element converted between one form and another is easy to see from the data. The main nitrogenous determinands are:

- Total Kjeldahl nitrogen — TKN (organic-N plus ammoniacal-N).
- Ammoniacal nitrogen — NH_3-N, or NH_4^+-N.
- Nitrite nitrogen — NO_2^--N.
- Nitrate nitrogen — NO_3^--N.
- Total oxidised nitrogen — TON (nitrate-N plus nitrite-N).

It is worth emphasising that an understanding of the nitrogen pathways and the various determinands – as described above – is essential if you are to design a sewage treatment system capable of removing more than just the BOD and SS.

Phosphorus

Phosphorus is a fascinating element in biochemistry. The reactions that generate energy in cells all involve compounds containing phosphorus. Whilst not directly toxic, excess phosphorus can be of concern in watercourses, which normally lack this essential nutrient. Phosphorus is the nutrient most critical to 'eutrophication' – over-richness of nutrients. The growth of, for example, algae in the receiving water is always limited by the lack of one nutrient or another. Often, phosphorus is that limiting nutrient, and when it is supplied the algae grow in abundance, blocking light, so preventing oxygenation by other plants, and sometimes producing toxins.

Half the phosphorus in domestic sewage is from our excrement; the end product of our metabolism. The other half is mainly the phosphates that are added to washing powders.

The main determinand for phosphorus is orthophosphate (usually referred to simply as phosphate – PO_4^{3-}), which is metabolically available phosphorus combined with four atoms of oxygen. Orthophosphate contrasts with pyrophosphate (two P atoms) and polyphosphates (several P atoms).

Where effluent is discharged to soil, groundwater or larger lowland rivers, phosphorus is of little concern. However, if a discharge is to a Scottish loch then phosphorus removal may be required even for a small system. Removal of phosphate is not simple in a wastewater system. It is possible to manipulate the amount of oxygen available to the sewage and create an alternation of aerobic and anaerobic conditions, which a bacterium known as Acinetobacter can then exploit, thus removing phosphate from the water. However this is so difficult and expensive to achieve, that it really has no place in the systems we might build for ourselves.

A more practicable means of phosphorus removal is by making it stick to something — a process called adsorption. (Adsorption means sticking to a surface, as opposed to absorption where something is taken into something else.) The main surfaces to which phosphorus tends to stick are rich in iron, aluminium, or calcium. Adsorption is possible when there are lots of stones or soil in the system being used. Clay in particular is full of aluminium and binds phosphate. Such adsorption systems are usually excellent when first used and then become progressively less adept because the phosphate binding sites become occupied. Indeed, with changes in pH the adsorption fluctuates and there can be releases of the previously bound phosphorus.

Some phosphate is also taken up directly into living cells and can be removed from the water by removing the organisms, either as sludge or by harvesting plants. A less 'green' but usually the most effective approach involves dosing with a chemical that brings the phosphorus out of solution — a two-stage process of flocculation and settlement. The disposal of this sludge then becomes the issue.

Sulphur
Sulphur is ubiquitous in small quantities throughout organic matter. We are not usually concerned with sulphur as a pollutant in our final effluent but sulphur chemistry may cause us problems of odour and corrosion in the treatment system. With aerobic breakdown, the main product is sulphate, which remains dissolved in the water and becomes sulphuric acid. This can attack concrete, so special cement mixes are required if high sulphur content liquids are being treated. In anaerobic conditions, metals too can react, giving a black coating to valves and fittings and one can smell the "rotten eggs" odour of hydrogen sulphide. Some municipal systems recirculate treated effluent high in nitrate to act as an oxygen source to counter odour problems in raw sewage pump wells.

Pathogens
Leaving analysis of the chemical components of wastewater, we now turn to biological determinands. Sewage carries a multitude of potentially pathogenic organisms (see Introduction), the removal or isolation of which is essential if water-borne disease is not to be transmitted.

How do we find out whether drinking or bathing water is contaminated with sewage-derived pathogens? We could look for the pathogens themselves, but they are present only in extremely low numbers, amongst a host of non-pathogenic bacteria. And although one cell may be enough to trigger a disease, one cell among billions is difficult to pick out. So instead, it is established practice to look for specimens of the billions of non-pathogenic (or at least not necessarily pathogenic) bacteria that are always present in human faeces. These bacteria are called indicator species.

The most common indicator species is Escherichia coli, members of this group being known as 'faecal coliforms'. Faecal coliform population density measured in units of cfu/100ml, in other words colony-forming units per 100 ml of water, is a major bacterial determinand for water. Faecal coliforms are rod-shaped bacteria that have certain unique characteristics (to do with the way they eat sugar at 44°C), so they are easily spotted. Furthermore, outside our bodies, they live almost exclusively in poo. So, if you find them in water, you know human or animal excrement has been in the water fairly recently; that human pathogens may also be present now; and that you should avoid drinking it. The legal drinking water standard says that even if there is only one faecal coliform bacterium cell in a quarter of a pint of water (in other words 1 cfu/100ml), you should not drink the water.

Of course, your sewage effluent will not be cleaned to the standard of drinking water!!!; and you will not be expected to achieve this. In fact in the UK a faecal coliform standard is (currently) not required for small private systems for effluent discharged to inland watercourses, because the risk to human health is considered to be minimal. Faecal bacteria are inevitably much reduced in number by the processes occurring alongside the removal of BOD and SS. It is presumed that anyone who drinks water from a river downstream from a sewage outlet will sterilise it first and that people don't swim in rivers (well, only a few of us!).

So what are the pathogen-killing processes occurring in your sewage treatment system? In general, the longer a pathogen finds itself removed from a host organism, the more likely it is to perish. In liquid-sewage treatment systems processes fatal to pathogens include:

- Being eaten by bigger organisms (especially protozoa).
- Adverse physical conditions (such as too low a temperature, too much sunlight, or too high a pH).
- The presence of anti-bacterial compounds (produced by some plants, as well as by other microbes).
- Competition for food with other microorganisms.

A worked example

Figure 4.3 shows a set of results taken from various stages of a real sewage treatment system (figure 4.4). By applying your newly acquired understanding you should be able to tell something about the way this particular system is functioning.

Line 1 The first line shows the concentrations of determinands of the liquid collected at point 1 on figure 4.4. These amounts are typical of primary effluent. The BOD is 250 mg/l, which is average for septic tank outflow. As a rule of thumb, and we must prepare ourselves for many a rule of thumb here, almost half the total BOD of incoming sewage is retained in a standard septic tank. SS is a little bit low here since, as another rule of thumb, we would expect SS and BOD to be about equal in domestic sewage. The TOC (total organic carbon) can be taken on its face value as being

66 Choosing ecological sewage treatment

	Susp Solids	BOD +ATU	Amm N	Nitrite N	Phosph P	TON	TOC
Septic Tank	134	250	38.0	0.082	13.0	<0.5	68
Primary Reed Bed Effluent	86	113	26.0	6.7	11.5	10.0	28.5
Feed to Iris Bed	50	53	21.5	2.55	12.0	9.0	16.0
Mixed Reed Bed Effluent	6	8.5	6.9	0.029	8.0	3.4	7.1
Final Effluent	2	4.0	11.0	0.104	9.0	3.4	8.2

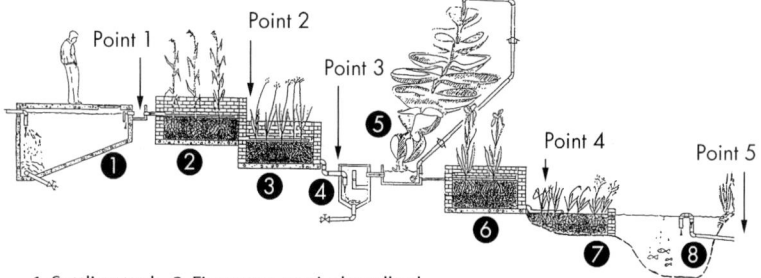

1. Settling tank 2. First stage vertical reedbed
3. Second stage vertical reedbed 4. Settling tank
5. Flowforms 6. Third stage vertical reedbed
7. Fourth stage horizontal reedbed 8. Pond

Figure 4.3. EA analysis of a small (reed bed) sewage treatment system.
Figure 4.4. The sampling points of the small sewage treatment system monitored.

Box 4.5. Water softening

Phosphorus is added to washing powders in order to reduce the influence of calcium and magnesium carbonate in the water, thus softening the water to make the soaps more effective. The irony is that calcium is sometimes added at sewage treatment works to flocculate the phosphorus out again!

68 mg/l and used simply as another comparison for the progression of the carbon transformation.

Looking at the determinands of nitrogen, we see that the majority of the N is in the form of ammonia, indicating that no nitrification has yet taken place and reinforcing the suspicion that this is septic tank effluent. Nitrite-N is very low as it almost always is, and TON is as low as the sensitivity of the test can measure. So we can say that there is almost no nitrate present.

The phosphate-P concentration is typical for a domestic situation.

Box 4.6. Best laid plans

Byron's Bay, in Australia, is a big resort for bronzing bodies and catching the surf. To protect the tourist industry it is important to keep the coastal waters clean and attractive. A big wetlands system has been added to the municipal works to make sure that only very clean water is returned to the sea. Reed beds of many types follow the aeration and flocculation tanks and are themselves followed by several hectares of marsh and heavily wooded ponds. The wetlands became a picnic site and people camped there overnight. However, the wildlife, which also gathered, was so abundant that occasionally the bird droppings increased the phosphorus level in the final outlet (admittedly from practically nothing to slightly measurable), above the level entering the wetlands!

Line 2 What do the figures at sample point 2 tell us? From a simple scan of the figures we notice that concentrations have all fallen except for the nitrite and TON. The rise has occurred because ammonia is beginning to be converted to nitrate. Notice that, although a great deal more of the nitrogen is in an oxidised form (TON), the total for [ammonia + TON] of 36 mg/l is very similar to the total of about 38 mg/l at sample point 1. It is tempting to assume that there has been a quantitative transfer of N from ammonia to an oxidised form of nitrogen, by nitrification. Indeed, as a general indication this is a valid assumption. However, experience has shown that it is likely that, within stage 2 of the treatment system, further ammonia is likely to have been produced from the degradation of organic-N. Knowing this, the results suggest that some N has been lost as ammonia gas and some lost as nitrogen gas via denitrification. However, the figures provided do not inform us of whether this has happened and to what extent. TKN analysis would have helped us to make this assertion with confidence. The nitrite figure is unusually high — an anomaly we must live with.

The phosphate-P measurement indicates a slight reduction through stage 2, probably occurring by adsorption to the sand and gravel of the reed bed, in this young system.

Line 3 No need to dwell on the BOD, SS and N figures of this line. All the figures indicate that the processes we would like to occur continue.

However, somewhat surprisingly we see that orthophosphate-P has risen again. Possible conclusions:

- Perhaps some polyphosphate has been degraded to orthophosphate.
- Perhaps the test is within its sensitivity and statistically the 'changes' are insignificant.

- Since the samples were taken within a few minutes of each other, the water which has arrived at the latter sampling point was predominantly from a washing machine, which used phosphates in the washing powder and the water higher up the system was from a sink, which has been released after washing down some vegetables.

We cannot readily distinguish between these possibilities from the figures provided. And that is another thing to get used to — dealing with probabilities.

Line 4 In the fourth line we see quite dramatic changes: everything is significantly reduced. Not only has *transformation* of the contaminants occurred, as witnessed by the BOD, SS, TOC and NH_4^+-N figures; but also their *removal* from the water stream, as indicated by the TON, and orthophosphate-P concentrations. We can conclude that both aerobic and anoxic zones have been encountered between the last two samples.

Line 5 The final water is of a high quality, far above usual legally required discharge standards. The trend of contaminant removal has continued up to sample point 5 but for one exception: the NH_4^+-N concentration has risen, and by an extent beyond the error margin of the tests. Perhaps further degradation of organic-N has occurred at this late stage in the treatment plant. Or perhaps sample 5 (which is not directly related to the foregoing samples since the water takes several days to pass through the system and all the samples were taken within 10 minutes of each other) was taken from a point that just happened to be carrying a relatively high level of ammonia produced some time previously and not yet removed. In fact, in this case, the latter is more likely since these samples were all taken at a natural treatment system that is made up of several consecutive stages ending in a pond. It is at this pond that sample five was taken. Knowing the system, it is possible that there was a reservoir of higher ammonia from the natural variation in treatment over the days. We can also conclude, from the fact that not all the ammonia has been removed, that the system has not been designed with sufficient capacity to achieve full nitrification of the sewage from the population served.

Biotic indices

One of the main reasons that we treat sewage is to protect the watercourse into which the sewage is discharged, so that 'higher' forms of life (such as fish) will be able to thrive in the water. Rearranging this line of reasoning, we should get an idea of the degree of health of a watercourse by looking at

GROUP	FAMILIES	SCORE
Mayflies	Siphlonuridae, Heptageniidae, Leptophlebiidae, Ephemerellidae, Potamanthidae, Ephemereidae	10
Stoneflies	Taeniopterygidae, Leuctridae, Capniidae, Perlodidae, Perlidae, Chloroperlidae	10
River Bug	Aphelocheiridae	10
Caddis	Phryganeidae, Molannidae, Beraeidae, Odontoceridae, Leptoceridae, Goeridae, Lepidostomatidae, Brachycentridae, Sericostomatidae	10
Crayfish	Astacidae	8
Dragonflies	Lestidae, Agriidae, Gomphidae, Cordulegasteridae, Aeshnidae, Corduliidae, Libellulidae	8
Mayflies	Caenidae	7
Stoneflies	Nemouridae	7
Caddis	Rhyacophilidae, Polycentropodidae, Limnephilidae	7
Snails	Neritidae, Viviparidae, Ancylidae	6
Caddis	Hydroptilidae	6
Mussels	Unionidae	6
Shrimps	Corophiidae, Gammaridae	6
Dragonfiles	Platycnemididae, Coenagriidae	6
Bugs	Mesoveliidae, Hydrometridae, Gerridae, Nepidae, Naucoridae, Notonectidae, Pleidae, Corixidae	5
Beetles	Haliplidae, Hygrobiidae, Dytiscidae, Gyrinidae, Hydrophilidae, Clambidae, Helodidae, Dryopidae, Elmidae, Chrysomelidae, Curculionidae	5
Caddis	Hydropsychidae	5
Craneflies/Blackflies	Tipulidae, Simuliidae	5
Flatworms	Planariidae, Dendrocoelidae	5
Mayflies	Baetidae	4
Alderflies	Sialidae	4
Leeches	Piscicolidae	4
Snails	Valvatidae, Hydrobiidae, Lymnaeidae, Physidae, Planorbidae	3
Cockles	Sphaeriidae	3
Leeches	Glossiphoniidae, Hirudidae, Erpobdellidae	3
Hog louse	Asellidae	3
Midges	Chironomidae	2
Worms	Oligochaeta (whole class)	1

Figure 4.5. The Biological Monitoring Working Party biotic index.

the creatures thriving there.

It is relatively easy to become familiar with what is living in the water into which you are proposing to discharge treated sewage, by sitting still for some time and observing the plants and animals – snails, larvae, newts, perhaps even fish – living there. A magnifying glass and a field guide can assist you (for example *Collins Photo Guide to Lakes, Rivers, Streams and Ponds of Britain and NW Europe*, ISBN 978-0002199995. Out of print but available via second hand on-line retailers).

By careful study it has been possible to gather enough information about the various life forms such that the presence or absence of particular species in an environment is indicative of the life-supporting capacity of that niche. This compilation of information has led to lists of organisms alongside their ecological significance. A table of such information is known as a biotic index.

One feature of using a biotic index, rather than biochemical methods (see above), for water quality assessment is that aquatic organisms are continuous indicators of water quality. These organisms are affected by discharges whether they occur when the regulator's pollution control officer sticks their sample bottle under the pipe or whether there just happens to be a little discharge of highly contaminated effluent in the middle of the night. There have been several attempts to formalise this general concept. The one we shall mention here is the BMWP system (figure 4.5). The Biological Monitoring Working Party was set up by the Department of Environment and National Water Council to make a score card (a 'green gauge') for the various invertebrates that might be found in freshwater habitats. These invertebrates are characterised in figure 4.6, which shows the typical effect of a sewage discharge on a clean stream in terms of the oxygen level and the range of invertebrates living there. Thus, if river bugs (Aphelocheiridae) and cray fish (Astacidae) are in the water, it is probably fairly high quality but if there are only midge larvae (Chironomidae) and worms (Oligochaeta), we recommend that you might think twice about having a long cool drink from the stream.

Invertebrates are used to judge the quality of water for a number of reasons: they are a diverse group that are practically ubiquitous in freshwater; they tend to stay put, which means the survey tends to be a fair reflection of what happens in the area and makes performing and analysing the surveys relatively cheap; and, because they are relatively long-lived, they reflect the standard of the water over a useful time span.

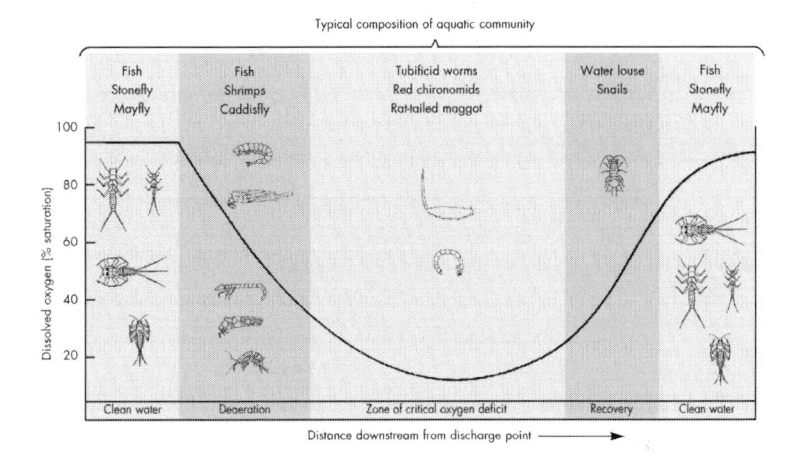

Figure. 4.6. Living determinands. Invertebrates as an indication of water quality either side of a sewage discharge point.

The weakness of most scoring systems is that they are not really standardised. To achieve standardisation would involve carefully recording the variety and number of creatures, taking into account the type of water body (for example lowland, clay pond or highland sandstone brook), and choosing the different micro-habitats (shaded or in full light, planted or unplanted) to sample at various points in the year and then averaging the results.

However, we shall leave all that to the professionals and give you a fairly simple method and a table for scoring. The method is known as the 'kick sweep sample'. One places a long-handled net (mesh size about 0.5mm, rigid frame) into the bank of the pond, river or stream while a friend holds a stopwatch. The best populated site is likely to be where there is some mud and shingle with plants rooted in it. For a given time (a minute is quite tiring) you thrust the net vigorously into the bank and lift up the material into the water. Repeat the process again and again until the timekeeper says stop. Then empty the muck, weeds, creatures, and so on from the net into a container, add some water, and take the container back to a place where the contents can be scrutinised beneath a lens.

A good field guide to invertebrates should be consulted, with care; a lot of these creatures look very similar to each other. After the critters are isolated and identified, they should be returned to the watercourse if possible.

So, you have a table of creatures and you have used your field guide to identify which family they belong to. Now add up your total to get your BMWP score. What did you score? 250? Incredible! 200 would be extraordinary.

The qualitative score alone (in other words not taking the quantity of each invertebrate into account) can now be compared to the RIVPACS score (River Invertebrate Prediction and Classification System). RIVPACS takes into account the width and depth of the water body, the altitude and distance from the source, the underlying substrates, the flow and slope of the river, and the calcium carbonate content of the water.

The RIVPACS score gives the best possible score for the characteristics of the watercourse, and the BMWP score gives the actual score based on your findings. The RIVPACS system can be found in a book called *The New Rivers and Wildlife Handbook* (ISBN 978-0903138703). If you divide the latter by the former (the actual by the potential maximum), you get the Environmental Quality Index (EQI). An EQI approaching one indicates there is no stress on the river/pond/stream and that all the invertebrates one would expect in such conditions are actually there. The lower the score, the greater the stress of the stream. EQIs are potentially comparable to the EA's (Environment Agency's — see Regulators and Regulations, below) river classification system ratings but this, as far as we are aware, has not been formalised.

Regulators and regulations

Regulations are in a constant state of flux and even established regulations are open to different interpretations; so individual regulators apply the regulations in different ways. Thus you can encounter (sometimes profound) differences in different localities. Partly to address this issue and partly as a cost-saving measure and to increase efficiency, the Environment Agency has recently centralised much of its operation. Main reception (08708 506 506) has trained staff to answer many questions. Beyond a certain point, however, it is still valuable to contact local officers, who have the insights more suited to your specific corner of the UK.

There remains a grey area at the interface of the designated roles of the EA and Building Control, so direct dialogue with the people who will deal with your site is recommended. What follows is offered in good faith as a guide. However our only *official* advice is to contact the local environmental regulator (EA, SEPA, EPA) and your local council's Building Control and Planning departments before installing a new treatment system or undertaking major changes to an existing system.

If you are building a new property or undertaking major renovation you will be contacting the planning department anyway and they will usually inform the relevant environmental regulator: the Environment Agency in England and Wales, Scottish Environmental Protection Agency (SEPA) in Scotland; or the Environmental Protection Agency (EPA) in Northern Ireland. You will have to show how sewage will be dealt with if no mains connection is available. The amount of detail required will depend on the size of development, the result of percolation tests and the location.

It is in your interest to make sure that any solution will be acceptable and early contact with the EA, SEPA or EPA is advisable.

Even if the work you are carrying out does not require planning permission any new drains or sewage treatment system will require Building Control approval.

Once detailed construction plans are submitted it is possible that the Building Control Department will require more details such as percolation results and calculations for leachfields or evidence that any innovative proposals (for example reed beds, trench arch) have been properly designed.

The relevant approved documents that support the Building Regulations are essential reading: Part H in England and Wales and Part M in Scotland. These are available for free download on the internet.

Environmental permits

Generally discharges from individual dwellings to ground of less than $2m^3$/day and discharges to surface waters of less than $5m^3$/d will not be issued a formal Environmental Permit (formerly 'Consent to Discharge') but you should always contact the relevant regulator to check for your specific situation (EA, SEPA, EPA). For these small discharges, provided they meet certain conditions (such as not being near a well or SSSI) a simple, free of charge registration with the EA is all that is required, exempting the discharge from requiring an EP. By law, all existing discharges of this type must be registered before 1st January 2013, after which time EP's will be required (although, at the time of writing (June 2012), the registration of small discharges in England is under review by DEFRA and the requirement may be abandoned altogether!).

Discharges of larger volumes, or from more than one property, require an Environmental Permit. Application for these is an involved process requiring the preparation of an "Environmental Management System" and a detailed Environmental Risk Assessment. There is also a hefty application fee and

> **Box 4.7. Royal commissions**
>
> Concern over water quality is not new. Early concern was expressed in Chadwick's 'Poor Laws' in 1850 and the Sewage Commission met in 1857. The most focused panel of concerned scientists who met to consider a strategy to deal with water pollution was called The Royal Commission. These panels put out nine reports between 1898 and 1915 and set a standard for discharge of polluted water to rivers. They recommended a standard of 20mg/l BOD and 30mg/l SS — hence 20:30 — into watercourses having an average flow rate of at least eight times the flow of the discharge. This figure held for a long time as the standard and is still known as the Royal Commission Standard. Nowadays, many situations require stricter targets.

an annual fee. Fortunately, for those few small discharges that *cannot* be registered and exempted, a simple EP application is allowed, with a more modest one-off application fee.

Whilst the regulator is not permitted to recommend a sewage treatment approach that is certain to be acceptable to them, they will need to be convinced that your proposal is likely to meet any standards that they may set. For smaller discharges, these standards may take the form of a 'descriptive consent'. For more significant discharges 'numerical' standards are set. Typically figures for suspended solids and BOD_5 will be quoted with 30:20 (see box 4.7) being the default for non-critical sites. More sensitive sites and larger discharges are increasingly likely to require an ammonia limit. If your discharge is to a low nutrient freshwater habitat, such as a loch, then nitrate and/or phosphorus removal may also be required. And effluent discharging to a beach or other recreational water is likely to require bacteria removal.

All the regulators provide clear guidance documents on their websites and are in general happy to be consulted.

Typically the system supplier or designer will deal with the regulator and help fill in any forms but if you have read this far then chances are you will be interested in following the process and understanding the requirements.

If in doubt, ask!

Environmental Health Department

Environmental Health Officers (EHOs) will not generally be asked to take an interest in your sewage system unless there is the possibility that a nuisance (for example foul odour) or health hazard may be created by the presence or absence of a treatment system. Usually the Planning Department will ask

Box 4.8. Analysis report

Results of a routine sample from a sewage system for 100 people.
(Reproduced with permission of the Environment Agency).

```
                              ANALYSIS REPORT
-----------------------------------------------------------------
SAMPLING POINT CODE 01039250           SAMPLE DATE:      11/11/98
                                       SAMPLE TIME:      12:45
SAMPLE FROM      OAKLANDS PARK         SAMPLE NUMBER:    394424
                 GRANGE VILLAGE        SAMPLE METHOD:    SPOT
                 NEWNHAM               SAMPLE REASON:    AUDIT
                 GLOS                  TAKEN BY:         610
                                       CHARGEABLE:       N
GRID REFERENCE: S068250941
DESCRIPTION: FINAL EFFLUENT MANSION HOUSE
SAMPLERS COMMENT:
-----------------------------------------------------------------
DETERMINAND                   RESULT VALUE    UNITS       CONSENT LIMITS
NITROGEN. TOTAL OXIDISED AS N   5.99          MG/L             :
PH                              7.6                            :
SOLIDS, SUSPENDED               6             MG/L    IM 0     :45
BOD + ATU (5 DAY)               <2            MG/L    IM 0     :25
CHLORIDE AS CL                  28.6          MG/L             :
AMMONIACAL NITROGEN AS N        7.78          MG/L             :
PHOSPHATE, ORTHO AS P           7.75          MG/L             :
BOD TUBE, FIELD                 <5            MG/L             :
WEATHER, AT TIME OF SAMPLING    DRY           MG/L             :
WEATHER, PREVIOUS 7 DAYS        DRY           MG/L             :
```

the Environmental Health Department to assess the situation if it seems necessary. This is most common when the system in question is unfamiliar (such as compost toilets or sewage treatment ponds).

EHOs have the power to make you take steps to meet their health, safety and nuisance concerns. They will not usually tell you what route you should take; it is up to you to submit a proposal that meets with their approval. That may mean obtaining clear evidence that your proposed system will work. Well established precedents are the most persuasive arguments in such cases, especially if the precedents exist within your district.

Summary

Assessment of the extent to which a sewage treatment system is achieving its aims is possible using a range of techniques. These vary from simple DIY tests to more complicated biological and biochemical laboratory methods used, for instance, by the regulatory authorities to assess and maintain standards.

Laboratory tests determine the concentrations of a number of 'determinands', the most frequently assessed being BOD and SS. By understanding the relevance of changes in concentrations of these, as well as other determinands, it is possible to assess the performance of a treatment system. Such understanding is useful for design.

Several regulatory authorities exist to offer advice, as well as to set and maintain standards pertaining to installation and modification of sewage treatment systems. Planners, Building Control, Environmental Health and environmental regulators refer to each other as appropriate, in assessing the most suitable way forward for each site. The growing familiarity of the authorities with novel treatment methods allows a wider choice of approaches to on-site disposal.

Chapter Five
Avoiding the Generation of Blackwater

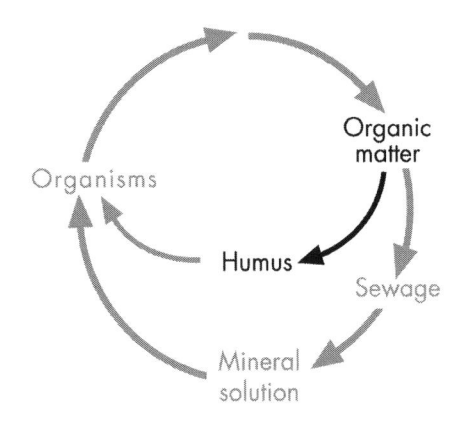

The Big Circle chapter described sewage as organic matter in water. This chapter considers whether it was necessarily such a good idea to add organic matter to water in the first place. The chapter is divided as follows:

1. The process of forming humus
2. How to make humanure
3. Returning humanure to the soil.

Soil versus water
In chapter two we saw that oxygen is central to the efficient breakdown of organic material. We might expect about 50% of the volume of a free-

draining topsoil to be occupied by air of which about 20% is oxygen. This is not the case in a body of still water, where even as little as 10mg of dissolved oxygen per litre (in other words per kg, which is 1 part in 100,000) is about the maximum possible concentration. Moreover, should that oxygen be used up (for example in sewage treatment), it would be replenished 10,000 times less readily than would happen through air. That is why, for instance, vertical flow reed beds are so much more efficient than horizontal flow.

This invites us to reconsider that the problem with sewage may not be so much that there is organic matter in the water, but that there is water around our shit! Because the water restricts sufficient oxygen reaching the organic stuff, it prevents a rapid and useful decomposition of our muck, leaving us with something that isn't particularly useful to man or beast.

So why do we use water? The reason is transportation and because water allows us to provide an effective, non-mechanical seal between us and where we are sending our wastes. Water moves the muck down the pipes and away from us so we don't have to deal with it right away. This has been fantastically successful from the standpoint of hygiene but by using water for transportation we turn two resources into one poison.

Humus

When organic matter decomposes on the surface of a lively soil some of the material is released to the air, particularly as carbon dioxide, due to microbial respiration. Another fraction of the organic matter is eaten by larger organisms that find the waste to be to their liking. These will eventually contribute their own droppings and carcasses to the same process. The remainder will be incorporated into the soil.

The part of the soil that is made of organic matter is called humus; and it's fascinating stuff. Humus has (at least) four important qualities:

- It contributes to soil structure.
- It provides energy.
- It is a relatively stable form of nutrients.
- It assists soil to interact with water.

These qualities are worth a little more scrutiny. There is a continuum of states of humus defined by two poles. One extreme is the persistent organic matter, which can be called gross structural humus. (Lawrence Hills calls it 'soil furniture'.) When it was in an organism it was also part of the organism's

Box 5.1. Colloids

'Colloid' is a term referring to one phase of matter (solid, liquid, or gas) dispersed within another. This differs from one phase dissolved within another. For instance liquid dispersed within a gas is fog, and solid dispersed in a gas is smoke. Solid dissolved in liquid is a solution (sugar crystals in tea), contrasted with solid dispersed in liquid which is called a gel. It is this last type of colloid with which we are primarily concerned when talking of the colloidal nature of humus and the smaller mineral particles. In doing so we follow a rather loose but widely used meaning of the term 'colloid'.

---

structure, as bones or lignins. The complementary pole is the colloidal humus (see box 5.1) formed either from rapidly decaying matter that once had its origin in the softer parts, such as the rapidly growing parts of plants, or the excreta or musculature of animals.

Topsoil is made up of this humus, along with the mineral elements of the soil, such as clay. Worms will blend these elements together to form small clusters, which give the soil a crumb-like structure full of air spaces. These spaces are also the passages through which water will travel as it makes its way through the soil.

For plants the colloids are the front line, the zone of activity and interaction between root, moisture, and nutrients. Generally the surfaces of these colloids are negatively charged, so they bind to positively charged ions such as potassium and magnesium. The positively charged ions can be exchanged for hydrogen ions from the plant roots, making the previously bound elements available for the plant. In this way too, humus acts as a reservoir of nutrients and minerals, preventing their escape in solution.

A soil rich in humus can hold much more water than the mineral elements of that soil could retain alone. What is important is that this water is able to exist alongside the humus without a significant leaching of the nutrients. Thus plant roots have a 'choice' whether to make use of clean water for transpiration or to absorb nutrients from the store of humus for their uptake.

This can be contrasted with discharging a mineral solution on to soil, whether this is derived from well treated sewage or from other water soluble fertilisers. In this situation, the majority of the soluble elements (particularly the nitrate) wash through the soil and have very little chance to add to the humus. Furthermore, plant roots in contact with the mineral solution are unable to avoid taking these salts into their tissues.

Box 5.2. Soak

Soak is usually a carbon-rich material such as sawdust, wood shavings or straw. Other materials can also be used, such as wood-ash or lime but, although these have the effect of smothering the scent of recent contributions and absorbing some of the moisture, they do not add carbon to the nitrogen-rich pile and they do not provide an open structure. For these reasons, when attempting to achieve ideal composting conditions, carbonaceous soak materials are preferred. Paper and cardboard are also OK, although they are not so good at smothering the dung and preventing smells. There is some indication that mixing the carbonaceous soak with soil actually reduces nutrient loss, by clay particles binding gaseous ammonia. So it is likely that the ideal soak is a mixture of chopped straw, soil and a touch of mature compost to seed the pile with organisms.

How to make humanure
Composting toilets

For human wastes the way to achieve the transformation from shit to humus is to use a composting toilet.

Compost toilets are not new-fangled gadgets; they are a balance between the ancient methods of the countryside (abandoned as inappropriate when urbanisation became prevalent) and the hygiene and convenience requirements of modern society.

What are compost toilets? At first glance, a compost toilet may look like a usual WC. But when you lift the lid of the pedestal (figure 5.1), the usual small pool of water is absent and in its place is a rather dark chute. You do what you came in to do and then, instead of flushing, you throw a handful of what is known as 'soak' (see box 5.2) down the chute from the container by the pedestal. Simply close the lid and leave the room with the knowledge that you have just made a small but significant step towards recycling your muck.

Figure 5.1. 'DOWMUS' ceramic composting toilet pedestal (Elemental Solutions).

What lies beneath the chute has a

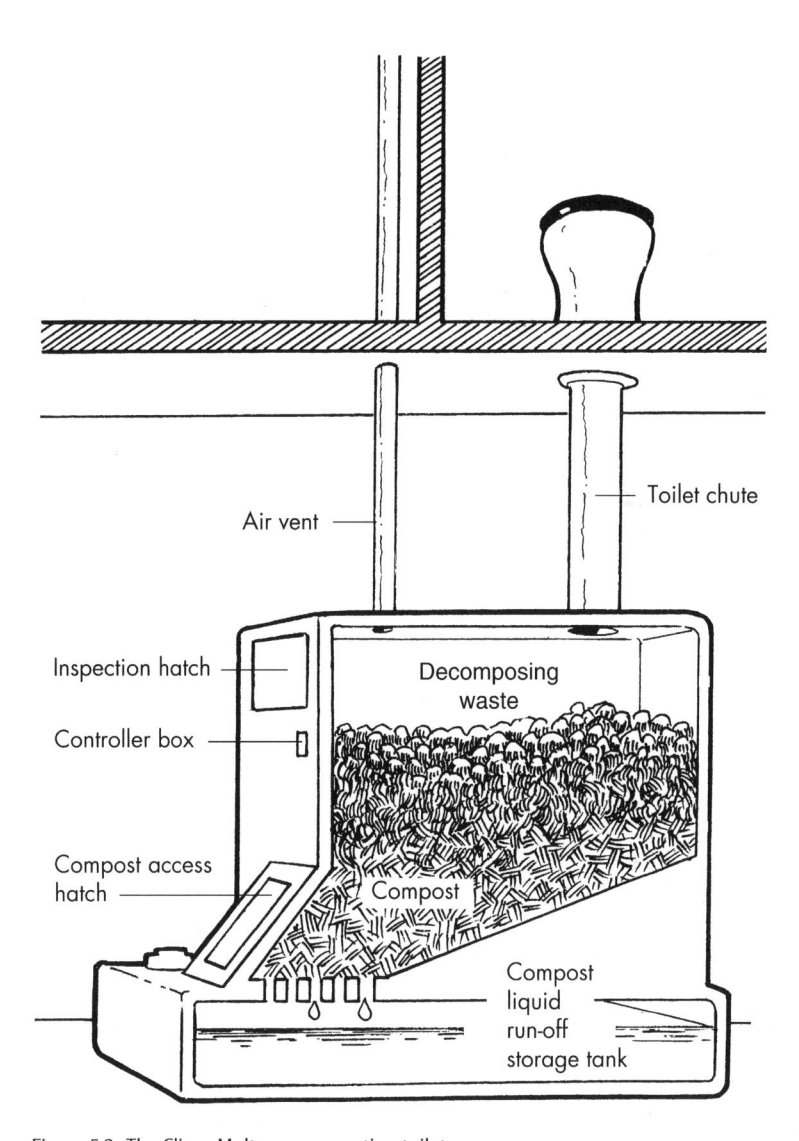

Figure 5.2. The Clivus Multrum composting toilet.

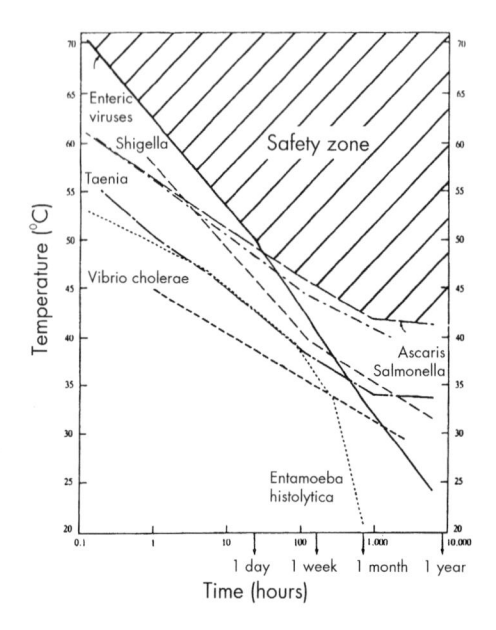

Figure 5.3. Pathogen destruction. The effects of time and temperature on human pathogen mortality. – Ater Gotaas W.H.O. 1956.
(Source: Gotaas, Harold B. *Composting: Sanitary Disposal and Reclamation of Organic Wastes.* WHO. Monograph series No. 31, Geneva, WHO. 1956. pg 35)

theme and variations. The theme is that there is a vented, drained chamber, which receives the faeces, urine and soak. The vent, which may be assisted by a fan, extracts any odour generated from the pile. The vent also takes air from the toilet room via the pedestal, thus removing the necessity for any other extractor fan. Apart from the drain, the chamber is closed in all other respects to exclude rodents and flies. The exit of the vent is usually covered by fly mesh to prevent the entry of flies, as well as their escape (thus limiting their action as disease vectors) should any manage to find their way in.

The chamber may be drained from the bottom or urine may be separated at the pedestal. Access is provided for removing the accumulated humanure, which is, if processed well, no longer recognisable as what you know it to have been.

Compost toilets and health
There are many pathogen-killing processes that occur in composting human manure on its way to becoming soil. Time is a major factor (figure 5.3) since

> **Box 5.3. Compost toilets and pathogen death**
>
> Clivus Multrum Inc, a company making compost toilets, has undertaken research into the material that leaves their toilets. Clivus has been awarded a seal under standard 41 of the National Sanitation Foundation scheme in the USA. Their work shows bacteria counts in their compost loos to be at least 10,000 times less than in sewage sludge from a septic tank, and a similar result was found for the liquid that drained from the installations, when compared to septic tank effluent. Indeed, the faecal coliform count was found to be lower than the recommended maximum for swimming quality water by an average of ten times! Current evidence indicates that it is advisable to leave the pile as untouched as possible for two years in the UK climate, or for seven years if *Ascaris* worms are known to be carried by people in your area, which is very unlikely in the UK. After this time the pile is pathogen-free.

the majority of human pathogens die within a few months once outside of the body. Exceptions include Ascaris lumbricoides and Strongyloides stercoralis, nematodes that can survive for several years outside the human body if conditions are optimal.

High temperature is another means of killing pathogenic organisms (figure 5.3). All human pathogens die if kept at 55°C for three days. But this is almost impossible to ensure. Since compost toilet piles rarely heat up (as a thermophilic compost heap in the garden would do) there is still concern about the efficacy of pathogen removal in compost loos. However, given enough time, even mesophilic (well below 55°C) composting is just about as effective at killing pathogens as the hot composting process (see box 5.3).

The challenges of composting human muck

Apart from the taboos (see below) and the health and safety aspects concerning humanure, (both of which we acknowledge and consider important) there are a few other potential problems with composting toilets.

Human turds can sometimes be surprisingly resistant to break-down, and excavated composting toilet containers sometimes reveal something very recognisable even after several years of hibernation. Although we have not seen authoritative research, the collective wisdom of the authors considers that there are three main reasons for this. The first is that with all the minerals, particularly from our urine, there can be very high salt and ammonia levels in the composting vessel. Worms and other degrading organisms cannot tolerate such an environment, so these essential degraders do not thrive. The second is that without sufficient soak and mechanical mixing (raking) the pile is likely to be anaerobic, creating a pickled preserve. The third reason is

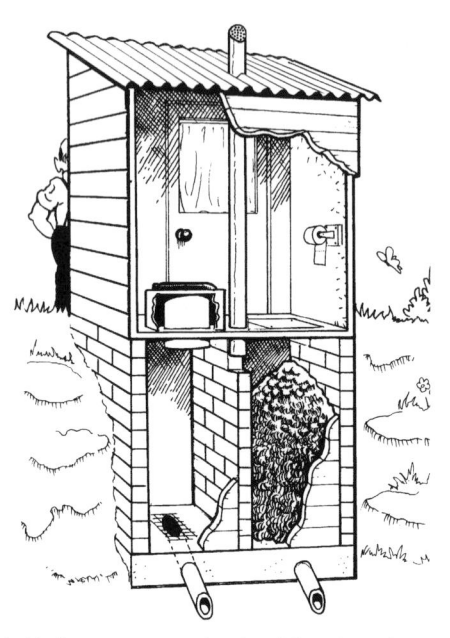

Figure 5.4. The Twin-Vault composting toilet. One full vault is left to compost for a year while the other is filled up. The first is then emptied out and the seat swapped over.

that the material is often too dry! Okay, we know we suggested not putting water into the toilet but for good composting some moisture is required.

For remote sites where sludge disposal is a problem but the convenience of conventional WCs is desired a Swedish invention came to our attention. This is the Aquatron (figure 5.5), which separates the solids (over 6mm) from the liquids in a flow of sewage, using the different properties of liquids and solids. It has no moving parts. It enables a flush toilet to be used and the composting process to be in a convenient position down and along the waste pipe. The Aquatron may be viewed as a way of providing the best aspects of the flush toilet with the best of treatment methods. As with compost toilets, the technology is a particularly good solution for sites without good access for the usual sludge tanker.

A final drawback of composting toilets is that they require some space beneath the pedestal, and this usually means in the floor below. For many people this floor below does not exist or is too precious to fill with a box of ordure! Composting toilet designers have developed some clever designs, which make use of moving parts to increase the space available without going down into the cellar.

Box 5.4. How much land do you need?

A fascinating study of what might be achieved with excrement has been conducted by John Beeby of Ecology Action, USA. Using simple mass balance calculations and experimentation on that basis, he has made and collated
meticulous research into the size, amount and makeup of the materials required for maximum retention of the 'fertiliser value' of humanure.

Beeby aimed for maximum reclamation of the nutrients in humanure by ensuring that the ideal carbon to nitrogen ratio (i.e. about 30:1) for a mesophilic humification process was present in his composting piles. He discovered that, owing to the low C/N ratio in humanure (see below), the astonishing amount of 15 cubic metres of carbonaceous material is required in order to retain most of the nutrients of just one person's annual production of faeces and urine. He then went on to work out the minimum land area required in order to grow this amount of carbonaceous material, and which would receive the humus produced to grow the food that would sustain the person producing the humanure in the first place. Dedicated work!

Joseph C Jenkins an independent researcher, also in the USA (who coined the term 'humanure'), has achieved a thermophilic composting process for humanure. The process loses more nutrients but has the benefit of quickly killing off the pathogens.

Both have published their work, Jenkins in his *Humanure Handbook* and Beeby in *Future Fertility*.

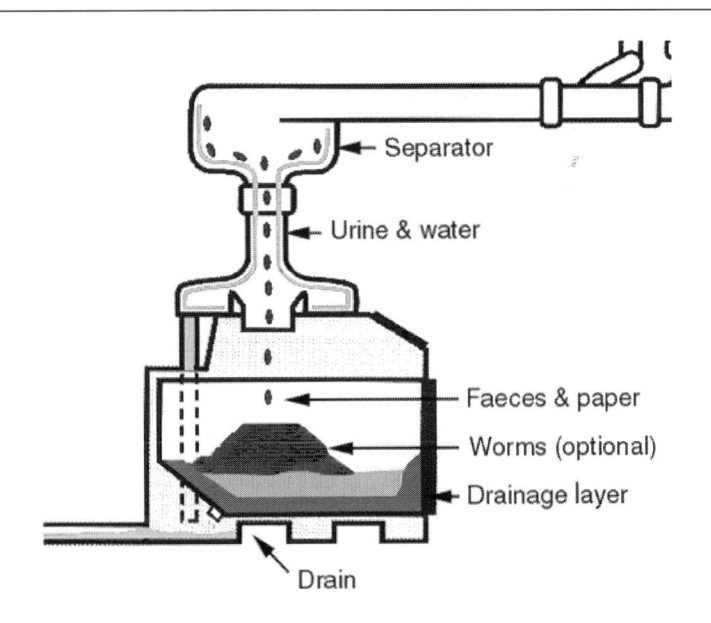

Figure 5.5. The Aquatron.

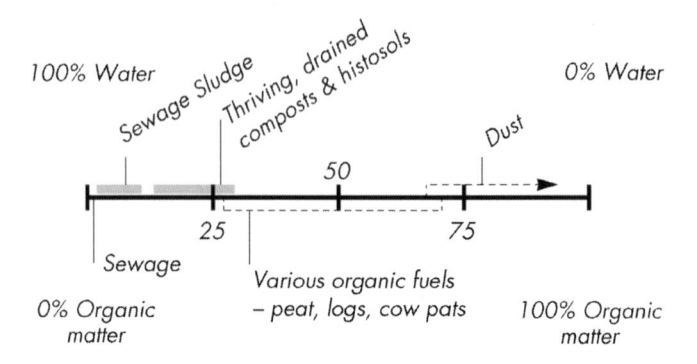

```
100% Water   Sewage Sludge   Thriving, drained                    0% Water
                             composts & histosols
                                                      Dust
                                         50
   ├────────┼─────────┼──────────┼──────────┼──────────┤──────▶
            25                              75
       Sewage
                       Various organic fuels
   0% Organic          – peat, logs, cow pats          100% Organic
     matter                                              matter
```

Figure 5.6. How much water? Graphic representation of the proportions of organic matter and water in various materials.

Urine collection

Separate collection of urine offers the benefits of both saving water and recovery of nutrients, since roughly three quarters of the nitrogen and half of the phosphorus we excrete is in our urine (see box 5.5). Fresh urine is generally sterile, making it relatively safe to handle. Collected urine can be diluted with at least ten parts water and used as a (water soluble!) plant feed, or added undiluted to a carbon-rich material such as cardboard or straw to make compost.

The simplest way to collect it is to keep a container in the loo with a suitable funnel and lid (figure 5.9 on page 90). A more refined way is to use a urine-separating toilet (figures 5.10.a & b on pages 90-91).

This clever appliance has a built in 'funnel' in the front of the normal toilet bowl to collect urine. The Ekologen (fig. 5.10.a) is flushed with only 0.1 – 0.2 litres of water. The urine is collected in a remote tank for use on compost heaps or as a liquid fertiliser and is adjustable. A solids-flushing model is also available requiring typically 5.0 – 7.0 litres per flush. Since the toilet relies on the user sitting on it for urine collection, a sloping ceiling or wide shelf may be needed to prevent men standing to pee, which could even keep the seat dry!

However such urine separating toilet designs are not without their problems and it is easy to imagine what can go wrong, especially when small children don't sit in exactly the right spot.

A far more robust solution has been developed by Andy warren of NatSol with early models tested at CAT. The NatSol urine separation system uses the coander effect to direct urine to a large gutter. Even if any stray solids, soak or paper sticks to the urine separator it will be washed off by urine and onto the composting pile, see figure 5.7.

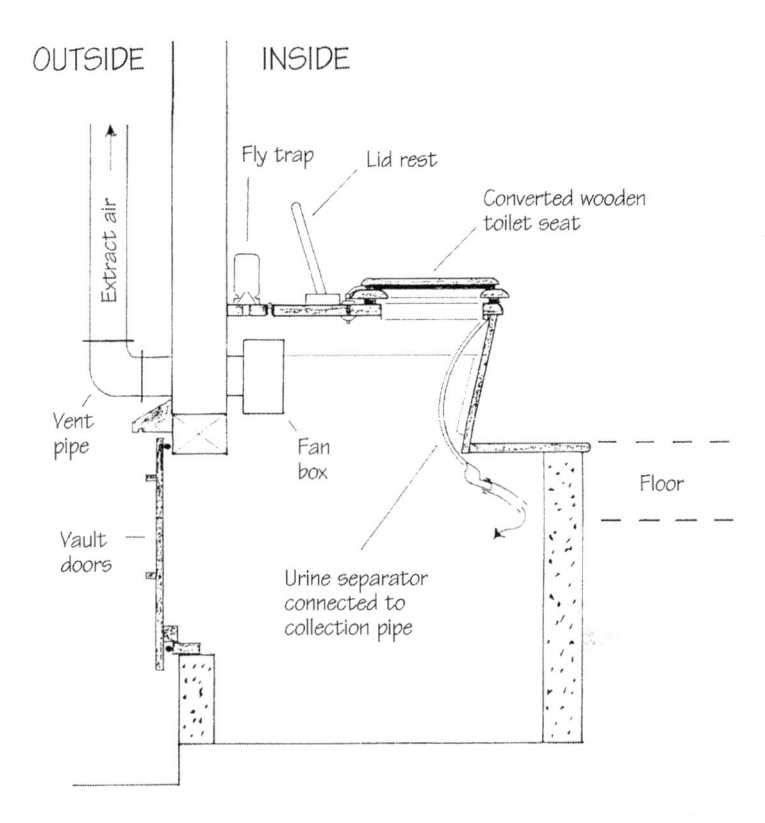

Figure. 5.7. Section drawing of Compus II. This is an old one-off now replaced by the NatSol product on the next page.

88 Choosing ecological sewage treatment

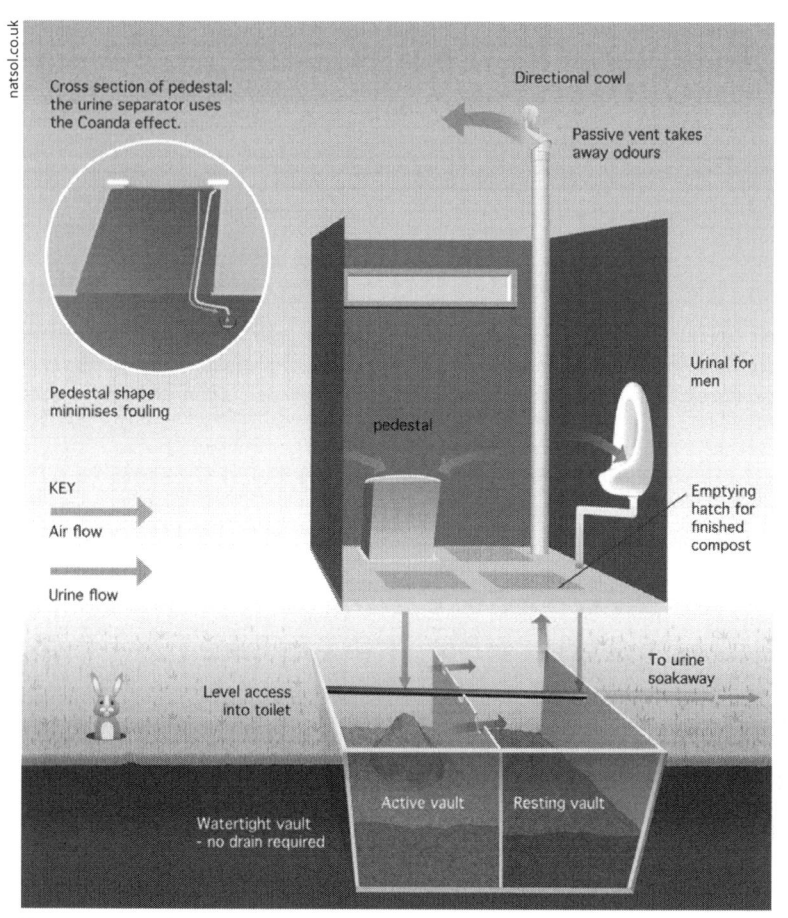

Figure 5.8. Natsol: how it works.

Having now completed our range of technologies by including composting toilets we can insert the Designer's Flowchart and the Designer's Checklist of System Types (figures 5.12 and 5.13 – see pages 92 & 94).

Electric toilets and chemical toilets

Whilst on the subject of avoiding making sewage by keeping the organic matter out of water, we should mention other toilets that don't have a flush.

Electric toilets (figure 5.11 on page 91) are toilets in which the excrement is heated to dry the matter to an inoffensive powder, which can be returned to the soil or put in the dustbin. They are often called compost toilets but the

Box 5.5. What is in your faeces and urine

	faeces	urine
Approx quantity:		
	135 - 270g per day (moist weight)	1 - 1.3 litres per day
	35 - 70g per day (dry weight)	50 - 70g per day (dry solids)
Approx composition:		
Moisture	66 - 80%	93 - 96%
% dry basis:		
Organic matter	88 - 97%	65 - 85%
Nitrogen	5.0 - 7.0%	15 - 19%
Phosphorus (as P_2O_5)	3.0 - 5.4%	2.5 - 5%
Potassium (as K_2O)	1.0 - 2.5%	3 - 4.5%
Carbon	40 - 55%	1 - 17%
Calcium (as CaO)	4 - 5%	4.5 - 6%
C/N	5 - 10:1	0.05:1
Total dry basis:		
N	3g (6% of 50g)	8.5g (17% of 50g)
P	2g (4% of 50g)	2g (4% of 50g)

Source: Gotaas, Harold B. *Composting: Sanitary Disposal and Reclamation of Organic Wastes.* WHO. Monograph series No. 31, Geneva, WHO. 1956. pg 35.

---

process is not 'composting' but 'cooking'.

These toilets successfully avoid plumbing, the contamination of water, and further treatment but since they are not really aiming to work within the Big Circle and because they use significant amounts of electricity we will not dwell upon them further.

Another non-flush approach is to store the matter in liquid chemicals and then throw out the chemicals and the matter within. These chemical loos are common, for instance in caravans and building sites, but their limitations are clear. They are there to avoid connection to mains sewerage and to keep the muck from smelling and attracting flies. But they just perform a holding exercise and do not work within the Big Circle.

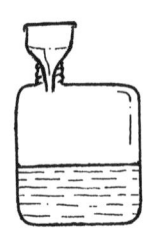

Figure 5.9 (top left). Back to basics: a pee can with lidded funnel will do the trick for collecting urine to use as a fertiliser.
Figure 5.10.a. More advanced: an Ekologen urine-separating compost toilet model (Elemental Solutions).

Returning humanure to the soil
A word of caution

The time will come to empty your composting chamber and do something with its contents. You may be surprised to find that, even with all the soak you added, the total volume is no more than a wheelbarrow of humanure per person per year.

You would be well advised to take all the appropriate hygiene precautions when handling the compost, even though it *may* be no more full of pathogens than normal soil.

So where should you put it? The best use of its soil-building potential is to work it into the topsoil with a hoe around the plants of your choice.

There has been much debate and argument about the health risk of using compost toilet contents on food crops. As there are no government guidelines for humanure reuse some officials have classified it together with sewage sludge, which does have restrictions concerning its application to the land. In fact this restriction is not pertinent, as the material does not contain the metals introduced by combining industrial wastewater with mains domestic sewage.

Avoiding the Generation of Blackwater   91

Figure 5.10.b. Urine-separating flush toilet by BB Innovation in a Swedish school. The urine is collected in large tanks and used in agriculture. Note the child-friendly lid (Elemental Solutions).

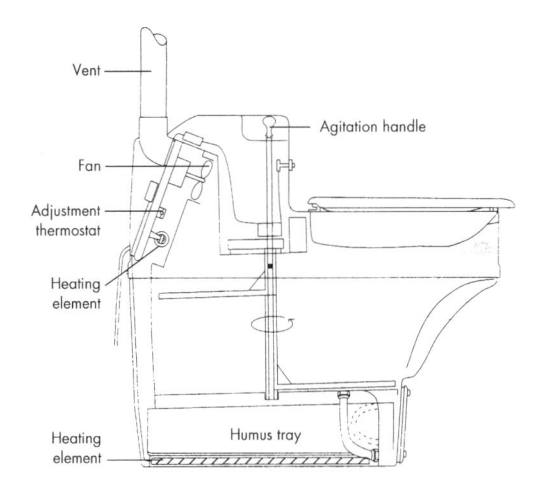

Figure 5.11. Cross-section of a Biolet electric toilet (courtesy of Wendage Pollution). Models exist using zero electricity or varying amounts, depending on the number of heating elements and the thermostat settings.

92 Choosing ecological sewage treatment

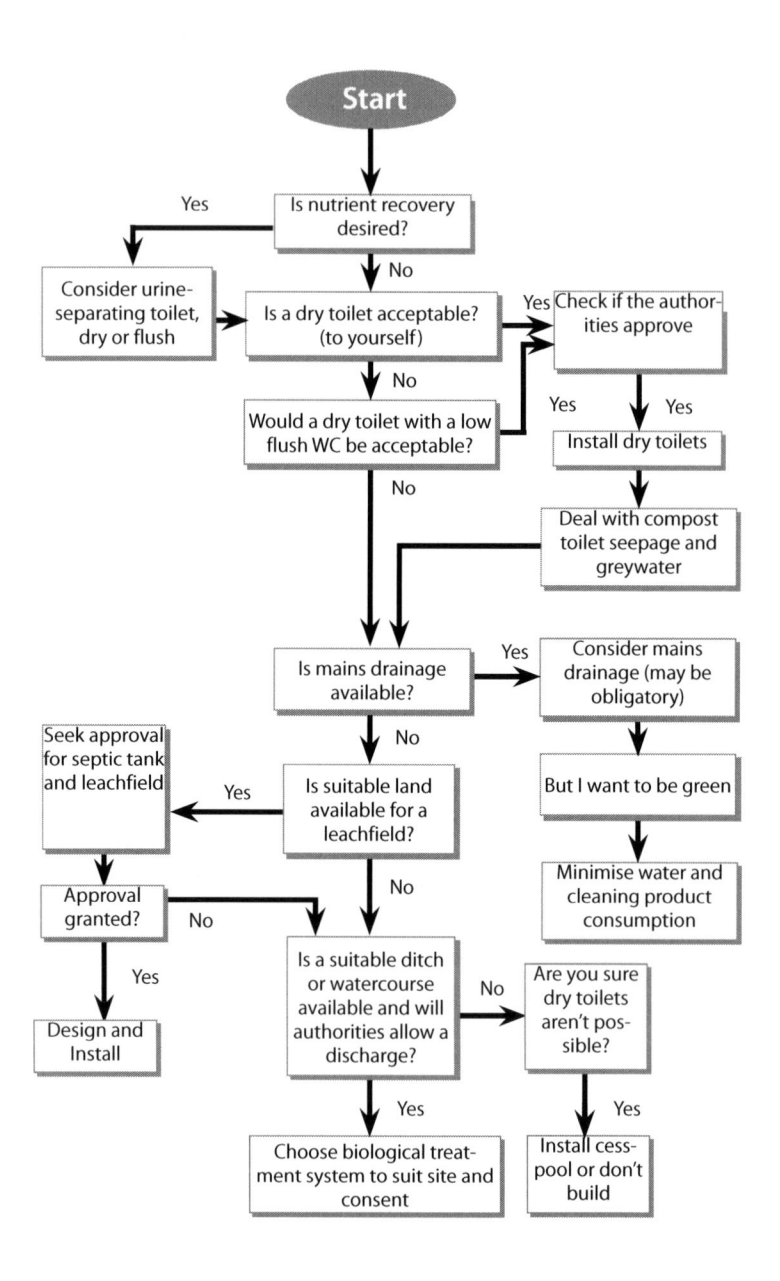

Figure 5.12. The designer's flowchart.

Box 5.6. Why separate urine from faeces?

Urine contains particularly high concentrations of nitrogen, phosphorus, salt and nutrients valuable to plants. All in a convenient liquid form, ready for use as a fertilizer, or for composting carbon-rich by-products such as cardboard or woody wastes (once diluted). But often more valuable to the dry toilet user is that without the wee the poo can compost more efficiently, with less mess and odour.

Wider discussion is respectful of other views and cultures, some of which recognise a taboo on the reuse of humanure, whilst others have it as a basis of their rural fertility.

So the authors' combined suggestion? In practice, the quantities produced from domestic systems are insignificant compared with the volume of farmyard manure or compost that a typical vegetable garden will require. We suggest the pragmatic measure of using well processed humanure for tree planting and ornamental gardening, after ensuring that drinking-water is not abstracted nearby or that there isn't a high water table, which might transfer pathogens or parasites to water supplies.

We recommend this because if we suggest that you draw your own conclusions it 'covers our backsides'. Ultimately it is a personal decision as to how you deal with humanure, from the extent to which you ensure hygiene in the toilet, to whether you apply mature humanure to your garden. However, the main reason (upon which we all agree) for dwelling on this is to reinforce a real sense of caution, essential simply in pure hygiene terms. It may be frustratingly slow to turn around the cultural momentum (currently in favour of water-borne sanitation) but there are real dangers in poorly handled muck. Sloppy work with compost toilets is bound to harm any potential progress towards their wider adoption. Any method of dealing with human waste should be undertaken carefully and any product that is for sale must be of a high standard.

The nightmare of the authors is that the eloquent persuasiveness of this book initiates a rash of stinking heaps of shit attended on all sides by flies and sickness.

In the hope of intercultural dialogue, and celebrating our common physiology, we leave this chapter with an anonymous irreverence.

Taoism	Shit happens
Confucianism	Confucius say, 'Shit happens'
Buddhism	If shit happens it really isn't shit
Zen	What is the sound of one shit happening?
Hinduism	This shit happened before

94 Choosing ecological sewage treatment

System	Land Use	Sludge Disposal	Effluent Quality	Energy Use excl. sludge disposal
Performance rating	Low — High	Low — High	Good — Poor	Low — High
Dry Toilets	May be difficult to retrofit	Dry compost low volume	None or very little	Low unless heated
Performance rating				
Cesspool	Access needed	Expensive as no sludge separation	None on site	
Performance rating				
Septic tank and leachfield	Land above field can still be used	Removal by tanker every one to five years	Very good	
Performance rating				
Package plant	Buried	Usually yearly by tanker	Medium	Electricity
Performance rating				
APTS*		Sludge drying bed often included	Good	
Performance rating				
Ponds		Infrequent but depends on design and climate	Medium to good	None to high depending on aeration
Performance rating				
Horizontal flow reed bed		Settlement or septic tank, remove by tanker	Poor to medium unless tertiary stage; then very good	
Performance rating				
Trickling filters		Settlement or septic tank	Medium to good	
Performance rating				

Figure 5.13. The designer's checklist of system types.

* Aquatic plant treatment system including vertical and horizontal flow reed beds.

Capital Costs	Odour and fly nuisance	Maintenance	Ruggedness	Aesthetics
Low — High	Low — High	Good — Poor	Low — High	Low — High
Low if self build, no pipes etc.	Little with proper design	Basic and can involve shovelling	Depends on designs but needs awareness	Can be very beautiful or a smelly Hole
Medium	Good	Emptying	100% reliable until full!	Underground
Medium	Good unless leach-field has failed	Preventative maintenance could save leachfield	Good	Underground
High	Generally good	Service contract common	Fair	Usually underground
High unless self-build	Fair, occasional smell possible	Simple gardening & ongoing awareness	Good. When it fails, does so slowly	Messy to beautiful
High unless self build but depends on site	Fair to good	Gardening	When it fails, does so slowly	Messy to beautiful
Medium	Fair to good	Low	Good	Messy to beautiful
Medium to High	Fair	Medium	Fair	Looks like a sewage system

Islam	If shit happens, it is the will of Allah
Quakerism	Shit happens to everyone
Catholicism	If shit happens, you deserve it
Presbyterianism	We never talk about things like that
Judaism	Why does shit always happen to us?
Atheism	I don't believe this shit
Apathy	I don't give a shit
TV evangelism	Send money or shit will happen
Rastafarianism	Smoke that shit
Protestantism	Shit happens because you don't work enough
Hare Krishna	Shit happens, rama rama
Jehovah's Witnesses	Knock, knock – shit happens
Optimism	Good shit happens
Agnosticism	Shit may happen
Existentialism	What is this shit, anyway?
Creationism	We just made up this shit.

Summary

Compost toilets are, potentially, the most benign method of on-site disposal. They avoid the use of water for transport and minimise sewage generation, treatment infrastructure, and pipework to and from the toilet. The humus generated is a stable, organic material with many features beneficial to soil and plants. However, care and caution are required.

Humanure (humus generated from composted human faeces and urine) can be produced using a modern compost toilet. Many toilet designs exist, some allowing separate collection of urine (which carries the bulk of the plant nutrients found in our bodily wastes) and others achieving simultaneous treatment of household wastewater.

For broad-scale agricultural use of sewage sludge other issues, outside the scope of this book, including Biodynamic and Organic certification, run-off, metals, and so on, need to be considered.

Great caution is advised when using humanure, especially in relation to food crops.

1. It is risky to use it on vegetables and generally not worthwhile at household scale because of the small volumes involved.
2. Putting it in the ground for tree planting and shrubs is the safest solution.
3. If you want to recycle nutrients then use diluted urine.

Chapter Six
Using Domestic Water Wisely

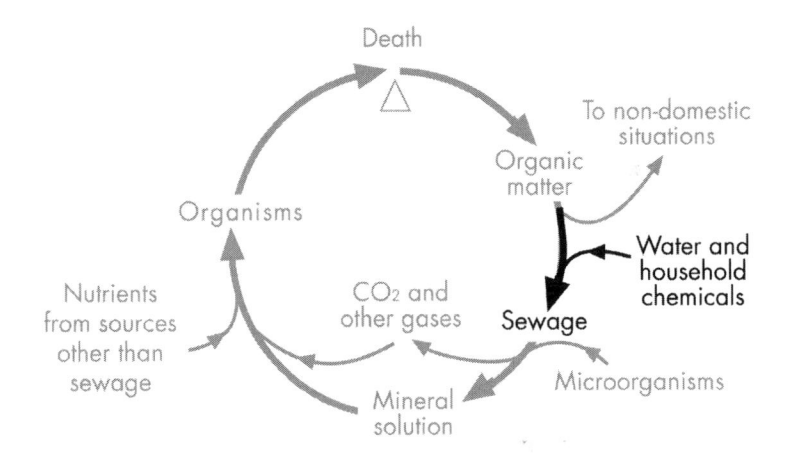

When it comes to making a sewage treatment system work, nothing is more effective than controlling what goes down the drain. The preceding chapter offered one approach to those who can, or need to, install dry toilets. The remaining water in the home can also be used wisely. That is the concern of this chapter, which is arranged in two main sections:

1. What you can put down the drain.
2. Water conservation in the home.

What you can put down the drain

Perhaps you have an existing treatment plant that is under-performing; or a leachfield that is getting wet and smelly and you're wondering if the problem is what you are putting down the drain. Or perhaps you have built your sewage system and you don't want to kill it by pouring harmful chemicals down the sink.

Clearly if you don't put anything down the drain your troubles will be over! Whilst this may seem a trivial observation, a radical reduction of your discharge volume can be a very cost-effective approach, particularly if you have a cesspool or where space for treatment and disposal is limited.

Detergents and household cleaning agents

Two common questions asked by people with their own sewage treatment system are, 'What detergent should I use?' and 'Is it OK to use bleach to clean the toilet?' The answer to both questions is that all of the systems discussed are robust enough to handle normal amounts of household detergents and cleaners. A clogged up leachfield will not be cured by a change of detergent brand. If you are concerned about the wider ecological impact of your actions, you will want to use the most ecologically benign cleaners in the smallest effective dose. We would have liked to offer a top ten rating but there are no clear answers. Most green claims found on detergent packages are meaningless. 'Biodegradable' simply means legal for sale in the UK and 'phosphate-free' washing up liquid is misleading since it is only clothes washing powders that might contain phosphates. Similarly 'septic tank friendly' means nothing.

One problem is that many different factors have to be compared and we struggle to equate transport energy with biodegradability with pollution at the factory. Nevertheless, we can offer you some pointers (box 6.1).

As an example consider the 'zeolite versus phosphate' debate. Phosphate is added to most washing powders as a 'builder', to regulate pH and improve the action of other ingredients, as well as preventing dirt from resettling on the clothes. Phosphate in our waterways is partly responsible for eutrophication (see chapter four) and the formation of algal blooms. Since around half the phosphate in sewage is from washing powders it would seem sensible to use phosphate-free powders.

The usual alternative to phosphate is an aluminium-based naturally-occurring, mined material called zeolite A, which has to be used with a 'co-builder' called polycarboxylic acid (PCA), which deals with the dirt resettlement problem.

A report by Bryn Jones (ex-Greenpeace) sponsored by Albright and Wilson (who manufacture phosphate-based detergents) gave the phosphate-based builder an impact score of 107 penalty points and the zeolite builder a score of 110. Whilst this difference is quite insignificant, Jones claims that 30 to 40 points could be knocked off the phosphate score by introducing

Box 6.1. Cleaning agent elements

1. Biodegradability: by law, all detergents are now at least 80% 'biodegradable' according to a recognised standard (OECD test) but some are more biodegradable than others.
2. Bleaches: washing powders, toilet cleaners and dishwasher powders usually contain chlorine-based bleach. The chlorine can combine with organic compounds to form highly toxic, and carcinogenic, organo-chlorine compounds. Non- bleaches are available and washing powders with separate bleach are a good choice (for several reasons).
3. Phosphates: phosphates are added to washing powders (but not to washing-up liquid) to soften the water. They can lead to eutrophication in watercourses.
4. Optical brighteners: these substances do nothing for cleanliness but give an illusion of whiteness. Problems include allergic reactions, poor biodegradability and mutation and inhibition of bacteria in your treatment system.
5. Other additives: NTA, EDTA, enzymes, preservatives, colourings, synthetic fragrances, etc., are all suspect in terms of ecological impact in production and final disposal.
6. Zeolite: possibly a more benign replacement for phosphate. It is an inert mineral but it can can encourage algae problems in sea water and comes all the way from Australia.
7. Sodium: common salt (sodium chloride) is used as a thickener in washing-up liquid; and soaps and detergents contain sodium ions, which break down the structure of clay soils and so reduce their permeability. Can be a problem for leachfields and greywater irrigation.
8. Manufacture, packaging, transport: however ecologically sound the final product, it is of little benefit to the environment if the manufacturing process causes significant pollution.
9. Political: it is impossible to look at ecological impact without considering the wider issues of fair trade, employee conditions, and a company's aims and ethos. Many of these issues are open to personal opinion so we leave them to your own research.

phosphate removal and recycling at sewage works. To add to the debate, Italian researchers have found zeolite to be implicated in the formation of 'sea scums' of algae.

If any of this is close to the truth, the main conclusion is that the quantity of washing powder matters, more than the brand. A little of the best will be better than a lot of the worst. With detergents, enough is enough and more won't make the wash whiter.

So, a lot depends on individual situations. If you irrigate your garden with your wastewater, then use phosphate-based detergents and grow big plants! If your sewage system discharges into a particularly sensitive body of water such as a loch then use phosphate-free detergents and let your sewage system

Box 6.1a Eco balls!

Figure 6.1. Several manufacturers now sell plastic balls or disks containing magnets, zeolite, or other substances for putting in washing machines. It is suggested that these reduce or eliminate the need for washing powder, although manufacturers admit that stubborn stains have to be removed first. Testimonials suggest that people are clearly happy with the result; however it is unlikely that many people do a proper comparison. As a counter anecdote, a curious colleague soiled a white sheet with a range of foodstuffs then tore it into four equal strips. The washing machine was run empty to clear out any detergents. Next a square of sheet was washed with 'balls' and no detergent, then another square with water only on the same program. The process was repeated with detergent and no 'balls' and finally with 'balls' and detergent. The first two washes were still grubby and were indistinguishable from each other whilst the next two washes were clean but again indistinguishable from each other. At worst, it seems likely that far less detergent may be needed than previously thought, although this will depend on many factors including water hardness, wash temperature, washing machine design, fabric colour and type, degree of soiling and personal standards. More research is needed.

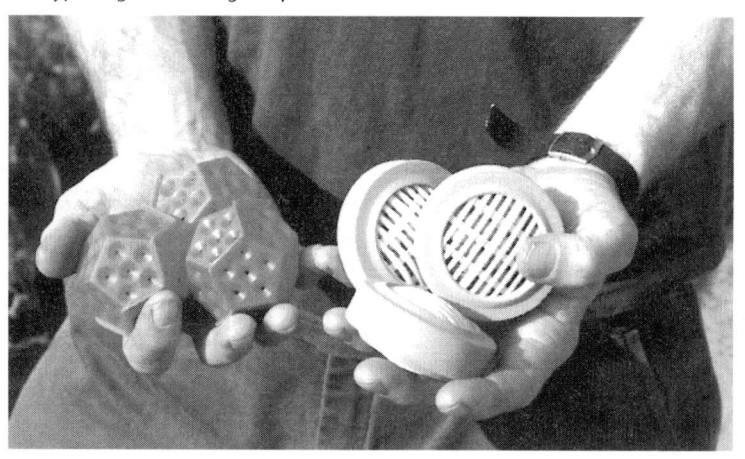

remove the non-soluble zeolite, ideally by discharging to land.

If all the apparent confusion has taken the wind out of your sails please don't be discouraged and please keep your ear to the ground for news of more ecological cleaning products, as there is clearly room for improvement, both in the products and the relatively new discipline of environmental impact assessment (EIA). Clearly some detergents are more biodegradable than others and contain fewer harmful additives, such as brighteners or artificial scents. It is just that we are at present unable to offer a simple recommendation with full confidence.

Box 6.2. Household solvents disposal

Solvent	Disposal method
White spirit	settle, decant and reuse, dispose of solid in bin
Brush washing water	pour on dry soil (small quantities only)
Engine/gear oil	ring oil bank helpline 0800 66 33 66 for local collection point for recycling
Old paint and garden chemicals	ring the Environment Agency helpline 0645 333 111 for your nearest recycling point
Cooking oil	compost, recycling centre for larger quantities
Unused medicines	take to the chemists
Rainwater	divert to rainwater soakaway or butt for irrigation

For general advice also phone the Environment Agency helpline 0645 333 111.

If the choice between pollution of the sea bed off Morocco (from phosphate mining) or open cast Australian bauxite mines (for zeolite manufacture) doesn't sit easily with you then all we can recommend is using as little as possible. There are books[1] listing household alternatives to commercial cleaning products. We have become accustomed to thinking that we need a cleaning product for every household task, when often all we need is water, which is totally benign — when used sparingly of course!

Other chemicals

Sheer convenience might tempt even the most conscientious of us to tip the occasional jar of white spirit down the drain when washing paintbrushes. Again small quantities should not kill your whole sewage system. The main concern is that such chemicals may receive little treatment and so end up in a watercourse where they could do harm.

One litre of white spirit is enough to bring 100 million litres of water above the recommended limit for drinking water! (The EC standard is 10 µg/litre for hydrocarbons.) This is the volume used by half a million people a day. If you use such things as white spirit then collect them for responsible disposal or recycling (see box 6.2).

Whilst, in an ideal world, we would all have a range of suitable containers for collecting these chemicals (or wouldn't use such chemicals), we recognise that very few people will take half a jar of brush cleaner to the Council's

---

1 *Clean Clothes and the Environment*, Jenny Allen, Women's Environmental Network, 1993

hazardous waste site. In this case, for occasional disposal of very small amounts of less noxious chemicals, the better of the possible evils is to sprinkle the offending liquid over a pile of soil or old compost. In the case of a liquid such as white spirit take the waste, leave it to settle, decant any clear liquid which may be reused, then allow the remaining liquid to evaporate and any residue can then be disposed of on a suitable site in the garden or in the dustbin. But do not use on the vegetable garden in case the solvent contains lead or other residues, and do not do this if near a well or other drinking water supply. Please note we do not recommend tipping chemicals in the garden but it is one step better than down the drain. Meanwhile, we hope that attitudes and infrastructure will change to make proper disposal the norm.

Water conservation in the home
Why conserve water?
Excess water can have a far greater impact on sewage treatment performance than the worst detergents and can lead to blocked or overloaded leachfields and reed beds. Perhaps counter to intuition, strong sewage is easier to treat than the equivalent organic load in a larger volume of water. Added to this biological factor, increased flows can lead to reduced settlement in septic and settlement tanks and so more solids are discharged to overload the next treatment stage.

The most serious problem is rainwater (from roofs and paved areas), which must not be put into the sewage system. 25mm of rain (an inch) falling on a 100m^2 roof equates to 2,500 litres of water. This is the equivalent of about fourteen people's daily water use concentrated in the space of perhaps an hour.

You can check whether your rainwater drains are connected to your sewage system by using buckets of water and, if necessary, food colouring or drain tracing dye (water is all that is usually necessary for domestic drains). Any offending drains must be directed to a separate soak pit in accordance with Building Regulations. Another potential problem is 'infiltration', where old drains have cracked and so let surface and/or ground-water into the sewage system. This can best be checked immediately after rain.

Reasons to save water:
- To reduce the volume of sewage to be treated.
- To reduce leachfield problems in heavy soil.

- To survive drought without standpipes or parched gardens.
- As a least-cost solution to meeting a growing need for water-related services, without flooding valleys or moving water by tanker or giant pipelines.
- To reduce demand on groundwater supplies.
- To reduce the threat to rivers caused by over-abstraction.
- To reduce the energy used to purify and pump it.
- To reduce the production and consumption of chemicals used to treat it.
- To reduce the energy used to heat hot water.
- To reduce water and sewerage bills.
- As a hobby and a challenge that doesn't need to satisfy strict payback periods.

Order of priorities: reduce, reuse, recycle

A general principle: it is almost always cheaper and more ecologically sound to save a resource than to recycle it or to harvest and transport more of it.

For example, a low energy light bulb costing, say, £5 will save more electricity per year than a £400 wind generator will generate in the same period. Turning off the light is even better.

Water use is no exception to this principle; it makes sense to apply conservation measures first.

Unfortunately there is little glamour in the message of turning off the tap while brushing your teeth or replacing the tap's washer when it drips. What we eco-philes want is a reed bed or rainwater system with pumps, pipes and valves to play with. The sequence below shows our recommended order of priorities for domestic water conservation. Here we are stating what we believe to be the obvious because it is not uncommon for the ecologically enthusiastic to go straight to number 6!

1. Develop awareness of use, and change to water-saving habits.
2. Fix leaks, dripping taps and cisterns.
3. Change or modify appliances starting with the biggest user (usually the WC).
4. Reuse water, without further treatment.
5. Harvest rainwater (for the garden at least).
6. Recycle greywater.

Let us look at each of these in a bit more detail.

Developing awareness, changing habits
The domestic water audit

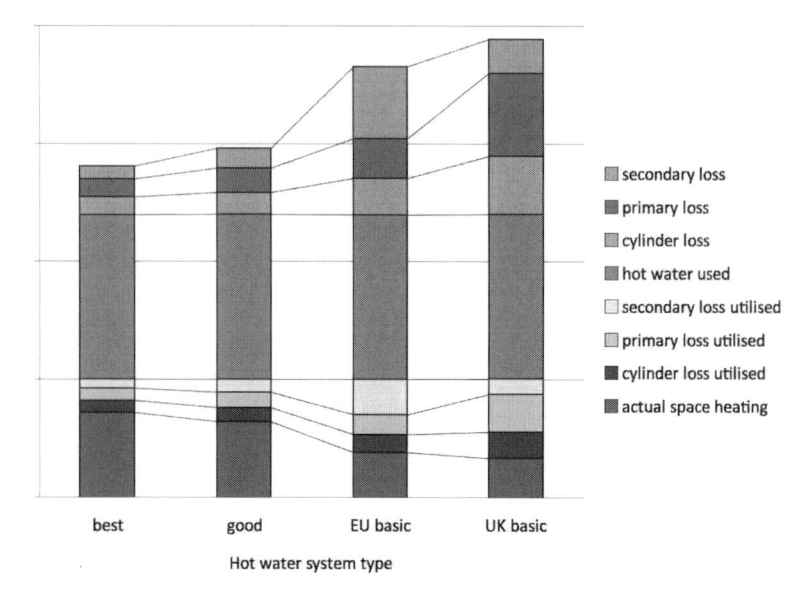

Figure. 6.1. Graph showing how hot water dominated heat requirements once we reach the Passivhaus level of energy efficiency.

Since water consumption depends on personal habits and variations in household plumbing, the first step to water conservation is to assess how much water is currently used, and for what purposes. Keep a note of how often washing machines and other appliances are used and estimate toilet flushes and shower use. Leave the plug in whilst you brush your teeth to see how much water you waste. If a water meter is fitted you can calculate how much water you use per person per day and check if your estimates add up to the measured consumption. This audit will identify the areas where most saving can be made and may show up some surprises.

Changing habits
Because most water is used in the bathroom and toilet, the biggest savings can be achieved by changes of personal habit, most of which are common sense. You may wish to take radical steps, such as not flushing the toilet after a pee or not washing. However, technical fixes are possible for those of you who want to keep your existing circle of friends.

Leak plugging
If you have a water meter then do a leak test. Read the meter then read it again after a number of hours when no water is being used. Even the smallest leak will be very significant as it will run 24 hours a day.

Although rather unglamorous, prompt replacement of tap washers or worn ball-cocks will save much wastage. A crude experiment with a jug and stopwatch demonstrates that even a slight drip can waste 30-40 litres/day (about 6 toilet flushes) and a steady dribble can easily waste water equivalent to having eight extra people living in the house — an enormous energy loss if it is a hot tap, and a real problem for many sewage systems and leachfields. If the WC flush diaphragm is worn or split there will be no leakage but the flush will be inefficient, leading to multiple flushing and wastage. If you are not familiar with simple plumbing repairs but are willing to have a go, then DIY stores often have free 'how to' leaflets.

Since 2001, WCs with a valve mechanism for flushing have been legal and are fast becoming the norm. Whilst the valve offers some advantages over the traditional UK siphon, for example the ability to have a button rather than a lever, such systems will fail sooner or later, leading to a waste of water unless repaired. Slow leaks can waste around 60 litres per day, yet be hard to spot. A regular check with a sheet of toilet paper held against the back of the bowl, above the water level, should show up even a minor seep.

Technical solutions — changing and modifying appliances
Some of these result in improved functionality — such as aerating taps that don't splash, WCs with an efficient flush, taps plumbed so that they run hot almost instantly, and foot- or sensor-operated valves to allow mucky hands to be washed without messing up the taps or risking contact with other people's germs. Here we shall discuss the following:

1. Toilets.
2. Showers, baths and basins.
3. Dead legs.
4. Combi boilers.
5. Taps.
6. White goods.
7. Water meters.
8. Gardens.

Toilets

Flush toilets traditionally account for around 30-50% of household water use, so the potential for savings is large. Obviously dry toilets (chapter five) give the greatest saving but, if you're committed to having a water closet, efficient toilets are available, designed to use around 4 litres per flush (figure 6.3a), and some existing units can be improved.

Valves versus siphons, dual versus single flush

Siphons were invented as a water waste preventer to solve the problem of leaking flush valves. Rather than a simple plug in a hole, the siphon cistern requires the cranking of a lever to lift some water and start the siphon. This arrangement cannot leak and is very robust. However in the original 1996 edition of this book, the authors questioned the fact that the UK was the only country in the world to require siphons and ban dual flush. At that time we saw the, then illegal, dual flush valves as the way forward.

By the second edition we had gained some experience and had gathered evidence from trials around the world and doubt about long-term water saving of valves was creeping in. When writing the third edition, valves and dual flush were not only legal but encouraged. Again, we are in a minority but now arguing for siphons or other leak-free robust technologies! A full discussion is outside the scope of this book but a number of papers and articles are available on Elemental Solutions' web site www.elementalsolutions.co.uk.

As far as flush volume is concerned the UK Water Regulations now demand a maximum of 6 litres. Unfortunately this is often insufficient, given the poor design of many pans and flush mechanisms. So, when purchasing, look for certification as evidence of good performance, such as WRAS approval.

Urinals

Whilst there is still the usual range of opinions, the authors are confident in stating that correctly specified and installed waterless urinals can outperform conventional urinals in terms of lack of odour and reduced drain blockage. Our own research is pushing for solutions that avoid the use of chemicals, throwaway parts or overpowering scents. Following the lead of the Building Research Establishment (BRE) we have simply turned off the water to several urinal installations and found that odours stopped. Whilst we are pleased to encourage independent experimentation, we should say that there are a number of details and caveats to observe. For example correct falls on pipes, appropriate traps and regular sluicing with warm soapy water to keep the traps clear. For situations where this level of maintenance cannot be guaranteed, two of the authors have developed the Airflush® which builds on

Figure. 6.2. The Airflush® waterless urinal. Courtesy of Green Building Store.

years of urinal and compost toilet experience. The Airflush was developed to address the numerous problems of flushed and waterless urinals with water-saving being almost incidental.

Improving what you have

Low flush toilets are a fairly expensive option if you already have a satisfactory loo. The simplest way to reduce the flush volume is to bend or adjust the ballcock. However this should not be done because it lowers the water level in the cistern and gives a poor flush. It is better to use plastic bottles filled with water, arranged carefully so as not to obstruct the ballcock and, if the overflow allows, to actually raise the water level by adjusting the ballcock. The trick is to determine how small a flush you can get away with — too low a volume will lead to multiple flushing and increased water use.

Whilst savings can be made by fine-tuning the flush volume, further savings can be achieved by the use of a smaller flush for pee, for which more flushes are required each day than for poo. Many siphon units (the plastic assembly inside the cistern) have the option of being used in the dual flush mode, usually by removing a small plastic plug (figure 6.3.b). If yours does

108 Choosing ecological sewage treatment

Removing the plastic plug allows air to enter and break siphon for half flush. Holding the flush lever down blocks the hole allowing full flush

Plunger lifted by flush lever

Plastic plug

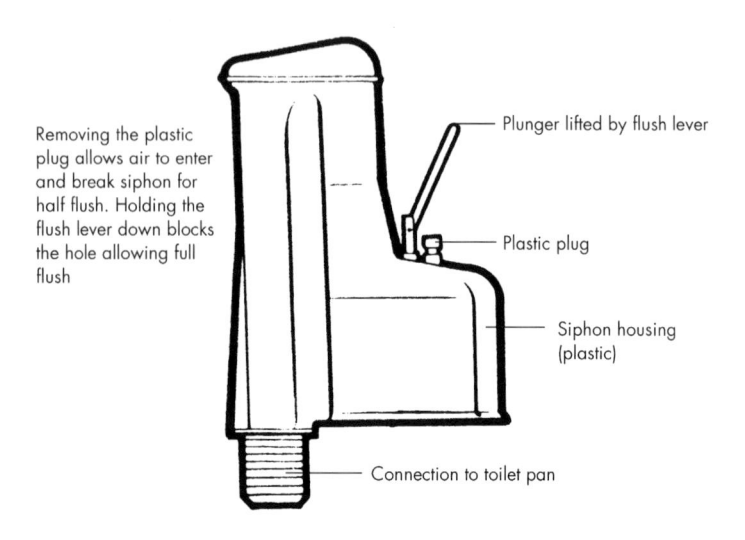

Siphon housing (plastic)

Connection to toilet pan

Figure 6.3. The Siphon.

Box 6.3. Siphon cistern or siphonic pan?

The siphon flush is the device described in the text and is a leak-free device for flushing WCs. The siphonic WC has a special pan that allows the trap and contents to be siphoned down the drain (a fault condition for a normal WC caused by poor drain design or blockage). The trap is then refilled with water to maintain a seal between house and drains. The siphonic pan is generally very quiet and effective in operation and may seem to use little water, as much of the flush usually bypasses the pan simply being used to start the siphon. It is our understanding that flush volumes much below 6 litres are difficult to achieve with available siphonic designs.

not have this, then kits are available to convert siphons to dual or interruptible flush. With the plug removed, the lever is pulled and released for a half flush or held down for a full flush. Since the instruction sticker provided is rather unappealing, it usually ends up in the bin, so visitors unfamiliar with such technology will probably waste gallons of water sussing it.

Retrofit dual flush siphons can typically save around 30% of WC-flushing water in domestic situations where users are familiar with the operation. For public toilets, savings are less and the risk of blockage with half flush is real.

Figure. 6.4. Four litre leak fee WC. Courtesy of Green Building Store.

Showers, baths and basins

Although showers are generally considered to use far less water than baths, consumer preference has pushed manufacturers to increase pressures and flow rates so that a modern power shower can, at a flow of 20-30 l/minute, use as much water as a bath (depending, of course, on how long you spend under there). Some full body showers with multiple spray heads use even more water with the limit being set by the size of the hot water cylinder. Low-flow shower-heads save water while maintaining the invigorating pressure but are usually designed to work at mains pressure. However, there are restrictors and flow regulators available that are easily fitted and will at least limit the maximum flow rate.

Many UK showers with electric heaters are, by necessity, low-flow, since the 7-8kW elements are able to heat only around 2.5-4 litres per minute to a reasonable temperature. 'Low-flow' is a relative term that seems to cover the range 1.5 to 10 litres per minute, so check actual flow rates and likely performance (as well as compatibility with your plumbing) before parting with your money. As a guide we would suggest 6 litres per minute as comfortable but not excessive and 9 litres per minute as an upper limit. With

Figure 6.5. The Opella Ecofil delayed action WC inlet valve. The WC cistern does not start to re-fill until the flush is finished thus saving water without reducing flush performance (courtesy of Opella).

good design and slightly lowered expectations a 4 litre per minute shower might be acceptable and still better than many older UK showers.

Dead legs

This is the length of pipe that the hot water has to travel through before it reaches the tap. An American website declares that letting taps run until water runs hot wastes 38,000 litres a year, which is over 100 litres per day in the average American household! (We have tried to find UK data.) The standard solution, found in hotels and some larger houses, is to use a pump to circulate hot water and so eliminate dead legs, but this incurs a significant heat loss. Complex systems with pumps, timers and temperature sensors to alleviate the problem exist in the US, but a simpler approach is possible for buildings without excessively long pipe runs. For mains pressure hot water systems (for example combi boilers), when installing sinks and hand basins, consider using microbore pipe (8 or 10mm copper or 10mm plastic) for the hot feed as it will run warm quicker (see box 6.5). Flow rate will be reduced, giving further savings, but check with a knowledgeable plumber or the pipe manufacturer before installing a long run to ensure the flow will be sufficient. In fact, where the flow rate to a small basin is to be reduced for water-saving reasons, this is best done by using the restriction offered by microbore, otherwise reduced flow rate will lead to a long wait until the water runs hot.

Luckily baths, which require a high flow rate, don't suffer in quite the same way from dead legs, as some cold water is usually needed.

Combi boilers

The main issue to be aware of with combi boilers is the water wasted whilst the heat exchanger warms up. A growing number of boilers now include a small well-insulated thermal store to eliminate this problem.

Figure 6.6. Ifö Cera 2/4 litre dual flush WC with drop valve flush mechanism. Courtesy of Green Building Store.

Taps
Hands-free tap
A nice technical fix for the enthusiast or gadget fan is a hands-free tap. Commercial taps with electronic hand sensors (figure 6.7) are available but are rather expensive at present. A simple switch-operated model is quite easy for a competent electrician to rig up. In the DIY model a switch activates a solenoid valve enabling hands off operation, thus reducing the temptation to leave the tap running while you brush your teeth or scrub your hands. You can use the small amount of movement in most under-sink cupboard doors with ball-catches to operate a microswitch, which switches on or off when you lean against the door (figure 6.8). Water savings aside, the fun to be had convincing visitors that your tap is voice activated makes payback calculations irrelevant.

Box 6.4. Dead leg volumes

Corresponding to Pipe Bore

Pipe bore (mm)	Bore volume per metre (ml)
6	18
8	33
10	56
15	133
22	340

[1000 ml = 1 litre.]

Low-flow taps and aerators

A simple trick for hand basins is partially closing the isolating valve that should be fitted on all recent plumbing fixtures, by law. This reduces the flow rate to the selected taps, saving waste as we rarely put the plug in when washing our hands or brushing our teeth. The potential problem with this trick is noise and the fact that flows may be too low when pressure is low. The solution is a flow regulator, which will maintain a constant flow over a range of pressures (greater than 1 Bar) without causing noise. The main advantage of fitting such devices is that flows are balanced, for example preventing temperature changes in the shower when the kitchen tap is turned full on.

Spray taps (figure 6.9) and aerators are designed to maximise the wetting capacity of a smaller flow and give the illusion of more water than is actually running. In the UK, taps incorporating aerators are available for use on mains pressure systems but will give a higher flow rate than typical UK taps, which often use a low pressure feed. Perhaps the biggest selling point of aerators at present is that they reduce splashing at equivalent flow rates, a common problem with mains pressure systems.

Clearly, low-flow taps on a bath would be counter-productive. Bath size and shape determine how much water you need to get a given depth, and insulation will help keep the water hot, saving on top-ups.

White goods

With the introduction of Energy Labelling it is now easier to compare the water and electricity consumption of new appliances such as washing machines. Machines that use less than 50 litres per wash cycle (compared with the normal 80-100 litres) are now the norm and cost no more. Reduced water consumption also means reduced energy consumption as less water has to be heated.

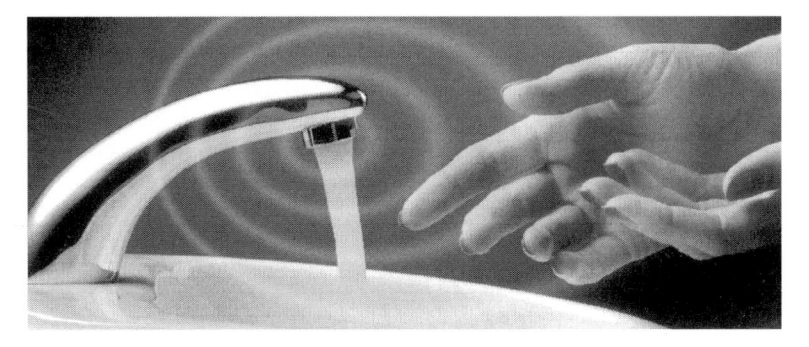

Figure 6.7. The Pulse 8 automatic tap detects changes in an electrical field (courtesy A&J Gummers Ltd).

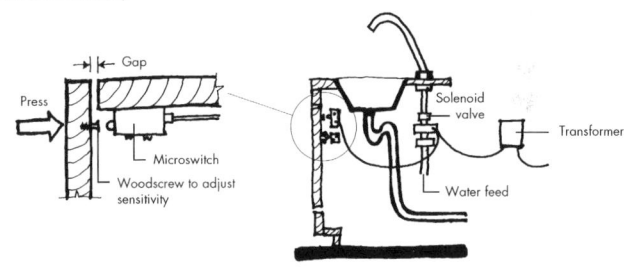

Figure 6.8. The DIY hands-free tap switch. Plan detail and side-view of door switch. The operator's knee pressing against a cupboard door beneath the sink can switch on the tap.

Water meters

There is little doubt that we are more careful when using anything that we pay for in proportion to the quantity we use. Water meters should allow the ecologically minded a small return on their investment in water conservation but the potential savings depend on family size, installation costs, standing charges and other tariff structures. Some see water meters as a tax on cleanliness.

Gardens

Whilst outdoor use typically accounts for only around 6 per cent of domestic consumption, gardens consume a lot of water at a time of year when it is in shortest supply. Water-saving gardening techniques range from simple mulching (but beware of slugs) to growing drought-resistant plants, to re-landscaping with clay linings under lawns. This is a complex subject as it involves living plants. For example, underwatering can be worse than no watering. See *Gardening Without Water* for more details.

Figure 6.9. This spray tap is activated by hand pressure on the top and will remain on for an adjustable period of 10 to 30 seconds (courtesy Pegler Ltd.).

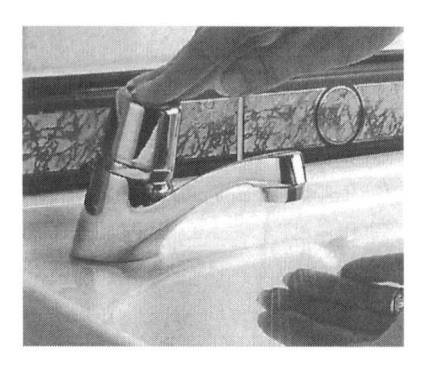

Reusing Water

Reuse involves using dirty water from one domestic activity for another less critical one, without treating the water in between. This is practicable only for greywater (see box 6.5). The most obvious and commonly practised reuse of bath and other washing water is for the garden in hot dry summers. Some caution is needed, however, as sodium salts from soap and detergents can build up in the soil and destroy its structure, particularly in clay. Gypsum can be added to the soil to help neutralise this effect. Special soaps are available, which use potassium – rather than sodium-based salts. If greywater is stored without treatment it will quickly go rancid and smell incredibly bad — even we won't touch it.

Rainwater harvesting

Harvesting the relatively clean water that falls on our roofs is a very old technology but the integration of rainwater into modern house plumbing is growing fast.

The importance and relevance of rainwater harvesting depends on the situation. In the UK most households are connected to mains water, so we are faced with the question of whether it makes ecological or economic sense for such houses to have a rainwater system beyond a garden butt. Certainly for domestic systems, the economics are far from attractive. If we consider a 50m^2 roof in a location with a fairly generous 800mm of rain per year we might at best expect to capture around 70% of this for use. 50m^2 x 0.8m x 0.7 = 28m^3/year. If we assume a generous water cost of £1.50/m^3 and an optimistic system cost of £2,000 installed we get a simple payback of 48 years, assuming no running costs such as electricity or pump failure. In addition and perhaps surprisingly, whilst electricity use for pumping is likely

Using Domestic Water Wisely 115

Box 6.5. Reusing greywater!

Sources in order of preference for ease of reuse

1. Shower and bath
2. Hand basins
3. Washing machine
4. Kitchen sink or dishwasher

to be modest, it will usually be significantly greater than the total energy to treat and deliver the equivalent volume of mains water.

Water recycling

It is possible to recycle household water but this typically requires chemicals and/or energy and involves potential health risks if the water is to be used for anything other than toilet flushing. Commercial systems are available for reusing 'light grey' water for WCs (in other words everything except the toilet and kitchen sink), increasingly via sophisticated 'membrane bio-reactors', but the authors remain unconvinced as to the practicality, economics or sustainability of current systems. While full recycling is technically possible,

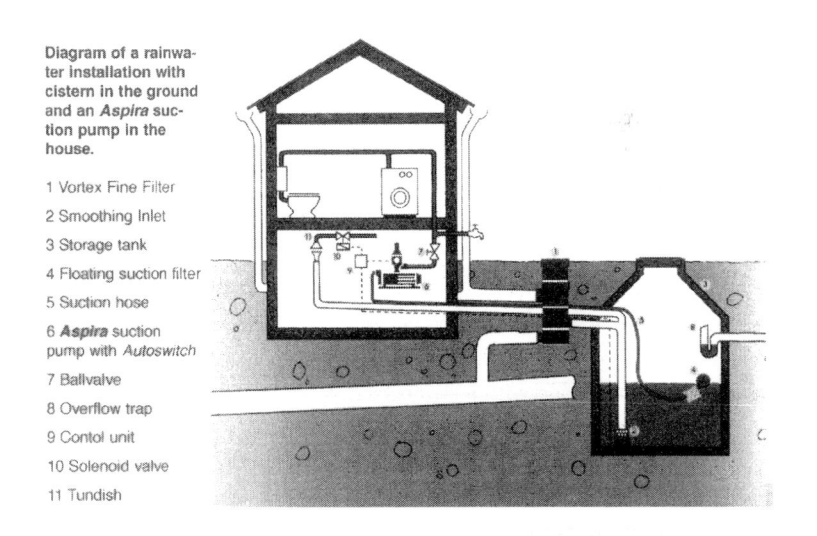

Diagram of a rainwater installation with cistern in the ground and an *Aspira* suction pump in the house.

1 Vortex Fine Filter
2 Smoothing Inlet
3 Storage tank
4 Floating suction filter
5 Suction hose
6 *Aspira* suction pump with *Autoswitch*
7 Ballvalve
8 Overflow trap
9 Contol unit
10 Solenoid valve
11 Tundish

Figure 6.9. Wisy® rainwater system for WC and washing machine.

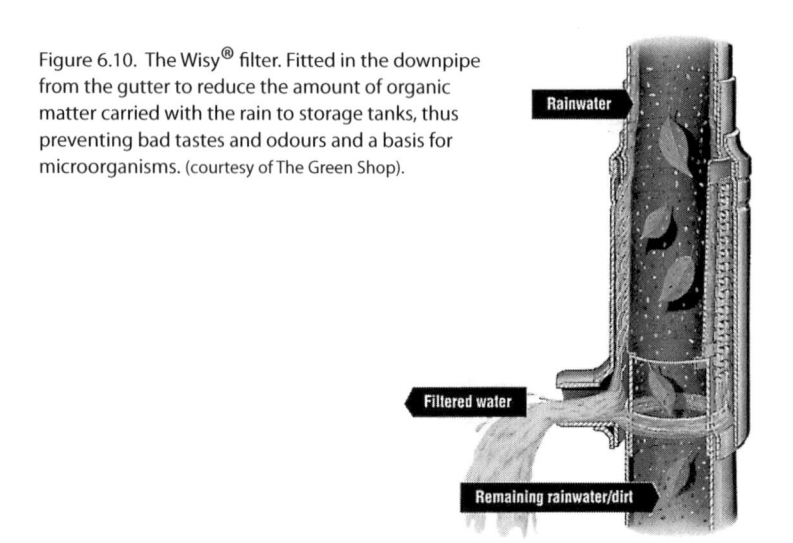

Figure 6.10. The Wisy® filter. Fitted in the downpipe from the gutter to reduce the amount of organic matter carried with the rain to storage tanks, thus preventing bad tastes and odours and a basis for microorganisms. (courtesy of The Green Shop).

we would argue that in the UK, if water conservation is the aim, then there is a lot to be said for letting nature broaden the circle. The natural water cycle can be seen as a solar-powered still, with remineralisation, filtration and storage on a vast scale. Our responsibility is to avoid stressing this system, either in terms of the volume of water we take from it or the quality of the water we return to it.

A correctly designed sewage treatment system (or even a simple septic tank and leachfield) discharging to the ground could be seen as a water recycling system replenishing groundwater.

Reasons for reusing or recycling wastewater:

- To reduce the volume of wastewater discharged, for example where a house is served by a cesspool.
- To water the garden in a dry climate or during a drought.
- To irrigate as a form of treatment and disposal.
- To use the heat from wastewater to warm a greenhouse.
- To reduce water supply and sewerage bills (if you're metered).

Summary

Appropriate use of your water supply and suitable disposal of greywater are major aspects of on-site disposal. The chemical constituents of domestic greywater are unlikely to damage a sewage treatment system. Water conservation can have major implications for successful sewage treatment and environmental impact. In order to minimise resource consumption, water-saving measures should be prioritised.

A range of technical suggestions and many appliances are available to help achieve water conservation.

Diverting greywater to the garden will reduce the load on the sewage treatment system or frequency of emptying of a cesspool.

For more information on all these subjects read CAT Publications' *Choosing Ecological Water Treatment and Supply* by Judith Thornton

Chapter Seven
Back to Life

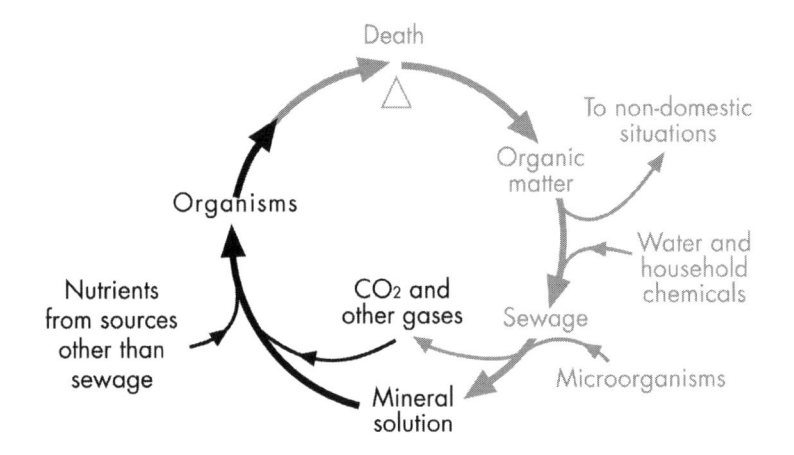

Diagram: cycle showing Organisms → Death → Organic matter → Sewage → Microorganisms → Mineral solution → CO₂ and other gases → Organisms. Additional inputs: Nutrients from sources other than sewage; Water and household chemicals. Additional output: To non-domestic situations.

After discussing sewage treatment and monitoring, we stepped back to explore how we can minimise our impact on the natural water environment. The final cleansing of our mineral solution we leave to the microorganisms, plants and animals of the natural environment but we raise the question of a possible human role here too.

This chapter considers the reincorporation of treated sewage into plants and animals of our choice, within the boundaries of our own land. It is organised as follows:

1. Humus
2. The Nutrient Solution
    Discharge to water
    Discharge to soil
    Aquaculture
    Irrigating willows
    Irrigating other plants

Humus

Before we focus on introducing our liquid effluent to selected organisms' environments, let us indulge in one last cheer for humus, the final product of the composting toilet, when it is the material we have to reintegrate into nature.

Humus:
- Is a stable form of nutrients.
- Is available to plants but not forced upon plants.
- Assists the soil structure.
- Increases the ability of soils to hold water in dry times.
- Increases the ability of soils to survive inundation in wet times.
- Stays put.
- Can exist with clean soil water without leaching of nutrients.

With that out of the way, and in acknowledgement that most people will have a liquid rather than humanure after treating their waste, we can look again at what might be done with the nutrient solution from sewage systems.

The nutrient solution

By one means or another, the sewage has been treated and flows to the end of your sewage treatment system. What remains is relatively clean water. As we have mentioned (chapter two), if your sewage has been treated to a secondary level there will be much that was originally in the sewage still present but it is now transformed into a mineral solution. If your system includes some tertiary treatment, some of the minerals from the solution have been removed. (If your system includes a pond or some plants, some of the water may have evaporated or been transpired.)

In bringing the effluent as close as possible to a mineral solution we have done a great deal to solve the problem of sewage. Can we now do something positive with the liquid that remains?

Discharge to water

If a particular site allows no positive use for the treated effluent, then it has to be simply thrown away. In many situations the only possibility will be to discharge the liquid directly to a body of water, such as the nearest stream, river or coastline. This direct discharge (in other words without a leachfield) must be done via a discrete pipe at the end of which the authorities can place

Box 7.1. Blue-green algae

In the late summer of 1989, in the UK, blooms of blue-green algae occurred, which killed some sheep and dogs and made ill a few soldiers who had been canoeing on the affected freshwater lake. The NRA quickly investigated the situation and put together a report. It concluded that some blue-greens exude poisons but it is not clear what factors determine why a bloom forms or why it releases poison. Conditions that produce a bloom in one lake do not always produce a bloom in another and it is no surprise that the report recommends further research and monitoring.

While other algae thrive in the spring when the majority of nutrients are available, blue-green algae have the ability to store phosphorus and thus outlive other species, blooming in late summer. Fertiliser runoff seems to be a relevant factor but algal blooms have occurred long before the invention of artificial fertilisers. Warnings about algal blooms are part of the culture of ancient peoples.

It is worth noting that in 1989, when blue-greens caused such a stir, 135 people drowned in reservoirs and canals but none died from blue-green algae poisoning.

---

a sampling vessel. This sampling point is arranged when you apply for an Environmental Permit (see chapter four). When you build the system be sure to leave access for the Environment Agency's pollution control officer.

Discharge to soil

In some situations, a flow of water with a loading of minerals is exactly what is wanted. Plants need nutrients and water to grow, and they need to take up nutrients in solution. Sewage effluent can provide this through direct irrigation to soil-based plants (or using hydroponic methods). At the same time, the soil and plants will help to treat the sewage by mechanisms that include:

- Oxidation of remaining BOD by soil microorganisms.
- Nitrification of remaining ammonia.
- Denitrification of nitrates.
- Bio-accumulation/absorption of all substances into biota.
- Precipitation and filtering of suspended solids and metals.
- Adsorption of minerals, particularly of phosphorus to iron, aluminium and calcium.
- Predation and ultraviolet destruction of pathogens.
- Cation exchange of metals.

Figure 7.2. Lemna harvesting on a grand scale (courtesy Lemna Corporation).

Of these mechanisms, perhaps the most important is the removal of phosphorus, which can be very thorough in clay-rich soils via adsorption to iron and aluminium. Too much phosphorus can cause algal blooms in receiving waters (box 7.1). However, the removal of phosphorus is not simple and is rarely achieved with any consistency by sewage systems. Generally, any phosphorus removal that does occur is a 'good thing'.

Ideally, treated sewage discharged to soil should be to the land surface (rather than underground). Odour is unlikely to be noticeable; once domestic sewage has been relieved of its organic load, only a very slight stale, sweet smell of treated sewage remains. It is important, however, to check with the authorities that discharge to soil surface is permissible in your particular situation.

Aquaculture

Another possibility is to practise aquaculture, which means growing animals and plants in water, usually for food (although you might like to cast a glance at the section on taboo [chapter five] before embarking). This is a very old use for nutrient-rich water and you can find more details in specialised texts[1]. However, you can establish a very crude version of aquaculture to polish your effluent by encouraging algal blooms to occur to your advantage. One simple way is to create an unshaded pond filled with sewage effluent in which algae are allowed to flourish. Sometimes these will be surface algae

1 *Plants for a Future* publishes a leaflet on edible bog and pond plants

that appear as a green tint or thin film. These will often be followed later by a billiard table finish on the surface of the pond (as when duckweed, for example Lemna spp, gets established). At still other times, the whole body of the water turns various hues of green like pea soup. At other times the pond may be dominated by filamentous algae, such as the blanket weeds, which seem to fill the pond with fibres.

This may not seem particularly advantageous but the algae initiate tertiary treatment of the wastewater. If one discharges this algae-rich water on to soil, a superb treatment system has been created. First of all, algae bloom in spring and summer when ambient temperatures are higher, so biological processes are faster and the BOD of the water is relatively low. This in turn means the ammoniacal nitrogen will be almost completely nitrified. Secondly, the soil receives the water with the minerals bundled into organic packets (the cells of the algae or Lemna plants) which can be filtered out much more readily by the soil than if the minerals had remained dissolved in water. The algae then decay on the soil surface where they can be incorporated into the soil as humus or, at worst, act as a slow-release soluble fertiliser.

This is now the basis for a type of treatment system and the Lemna Corporation, in the USA, has a wonderful machine for harvesting such plants from purpose-built lagoons (figure 7.2) receiving treated municipal sewage effluent.

However, if you're not going to discharge the algae-rich water to the soil or some kind of filter you have to be careful. If the discharge of the water is directly from the pond to a watercourse, the pollution control officer will need to take a sample of water and put it through the tests for water quality. These tests are not designed to distinguish between the various forms of solid or organic matter. So they cannot tell the difference between, for instance, the organic matter that was in the sewage before treatment and something relatively benign, such as algae, which has grown in the breakdown products of that original matter. Therefore, without careful design of the pond overflow features, you could receive a final discharge effluent analysis showing disappointingly high levels of SS and BOD, despite all your care.

Some countries, which are more familiar with pond sewage systems, make a 'chlorophyll allowance', recognising that, in a way, it is not fair to judge water full of algae in the same way that one would judge a sample with a turd in it! However, this is not the case in the UK. So if your final effluent breaches the legal consent limit you would be reprimanded whatever the nature of the organic matter in the pollution control officer's jar.

Figure 7.3. A well-grown willow plantation irrigated with treated sewage using the trench method (site system, CAT).

Irrigating willows

In practice, discharge to soil, with or without intervening ponds, has been made to areas planted with willow.

Two main strategies for distributing effluent to willows have been adopted by DIY sewage treatment boffins. One is to divide the total area earmarked for receiving the water into two or three smaller areas. Ideally, these areas are relatively flat and receive the water in succession. Every month or so a new section is put 'on-line' and the previously saturated section is rested. By alternating between saturation and rest, binding of the soil, which would restrict the permeation of the liquid, is avoided.

The second method of distribution is to make a channel, which meanders through the planted area (figure 7.3) allowing the water to ooze away through the banks of the channels and into the soil. The channel should fall at about 1:200 and so is appropriate for either level or sloping sites. Again, distribution can be alternated between one or more areas for the reasons mentioned.

Which willow strains you use will depend on what you want to do with them afterwards and on which ones will survive in the moist conditions they will meet. Commercial suppliers of willow can advise.

If you intend to grow a stand of willows for chipping and burning, one or more of the faster growing 'super willows' would be chosen – hybrids

> **Box 7.2. Seidel and antiseptic plants**
>
> One of the driving forces behind the development of reed beds came from the pioneering work of Käthe Seidel. One of her insights was that significant quantities of pathogenic bacteria died after passing through a horizontal reed bed planted with Schoenoplectus lacustris, the true bulrush. It was Seidel's opinion that the roots of the bulrush exuded an anti-microbial substance, although her evidence to support this is not compelling.

selected for vigorous growth, including Salix viminalis and S. dasyclados. And you will need to provide access for harvesting.

Basket makers might prefer some of the traditional varieties such as S. triandra, S. purpurea and S. viminalis. If you are growing the willow as a crop for craftwork, it usually needs to be clean and good quality so you'll have to make decisions about herbicides and fungicides.

The young trees, grown from cuttings about 30 cm long, establish best without competition for at least the first two years. We have used plastic mulches of leftover pond liner dug into the ground, with the cuttings planted through them (figure 3.13.a on page 47). Others have used bark mulches, which seem to work fine. Without mulching you will have to either accept much slower growth or return and weed the soil around the trees to make sure they get off to a good start.

Irrigating other plants

If you want to make a wildlife habitat, use a variety of species. We planted one plantation with willow, poplar and alder species, since they are known to tolerate saturated soils for part of their growing season. Oaks were planted into the banks, which divided the main planting areas, following a hint in the autobiography of Lawrence Hills, *Fighting Like the Flowers*. (The 'broken cycle' of fertility is something that also troubled him.) Oak is unbeatable at encouraging a wide variety of insect life and, therefore, increases the biodiversity of the whole area. (Willows are second only to oaks in terms of the number of insect species they can support.) Hills reported that the oak and walnut timber was of a high quality, even though the growth was greatly enhanced by being irrigated with sewage. One would expect that the 'forced growth' caused by the extra nutrients would produce a weak and sappy wood but this seems not to have been the case. We added ash (the tree, that is) a heavy feeder, along with anything else we could get our hands on, in the name of diversity. A fitting tribute to Lawrence Hills would be to use comfrey (bearing in mind his usual caveat of using a variety that will not spread).

One species that will no doubt volunteer even if it isn't planted (and will appear even if the water is treated to a very high degree beforehand) is the common nettle, Urtica dioica. Nettles are famous for gathering around muckheaps and septic tanks. They seem to favour soil rich with nutrients and bring the soil to a beautiful tilth around their roots. If they are happy there then they are probably appropriate, although we haven't read any research suggesting what their specific sewage-treating benefits might be.

There are so many beautiful and useful plants that like plenty of nutrients and thrive in moist ground that the reader is referred to more specialised literature.

Last word

We have chosen not to labour water's ecological crisis. Our agenda has been to elucidate the details of the problems and possible solutions of sewage pollution, in the hope that individuals will then be able to work better on their own corner of the situation.

It is beyond the scope of this book to give details of design and installation. Rather, we have offered an overview of the processes and technologies, as a first step to determining your own requirements and possibilities. We hope that by encapsulating this exposition within a grander context you will not only achieve a more informed position but also embrace the matter as a challenge, with inspiration and enthusiasm.

May you enjoy *Choosing* your *Ecological Sewage Treatment*!

(Now please wash your hands.)

Case Studies

We have tried to provide you with a basic understanding of on-site sewage treatment in order to allow you to make informed decisions regarding the optimum solution to your sewage problems. Here we show how theory can be put into practice with some examples of appropriate sewage treatment.

The following brief case studies are from our own work and feature, therefore, aquatic plant systems, and soil- or compost-based sewage treatment. These were deemed the most appropriate solutions for the given circumstances. It must be borne in mind, however, that in other situations, 'conventional' approaches for example standard package plants, septic tanks and leachfields, would be more suitable. The emphasis is always on the most appropriate approach for each given situation.

Case study 1 – Family house, South West England
Situation/problem
Very keen gardeners, with old septic tank and 'boggy smelly patch' in the lawn, wanted to sort this out and have water available for a pond in their orchard. Usually two people but occasional parties and visitors.

Solution installed
Septic tank fine so vertical flow reedbeds installed followed by a pond. Dosing device installed to ensure good distribution. Pond as final stage after humus settlement tank.

Notes/reasons
Beds made in manhandleable precast tanks for simplicity and to avoid trampling orchids and other treasured plants during construction. Reed beds particularly suitable for plant lovers as opposed to mechanical systems. Fall available in the garden. Good cleaning needed for pond to thrive. Pond overflows to make a bog garden, as desired, for wildlife and variety of plants.

Schema/sketch

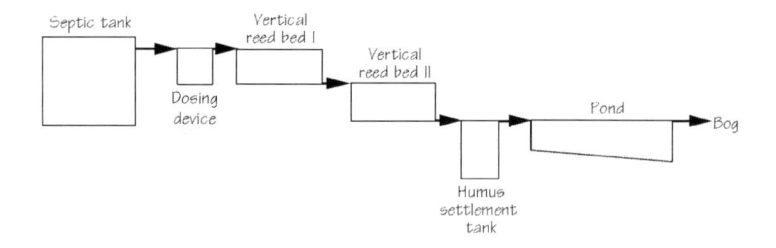

First vertical flow bed – just installed (Elemental Solutions).

A year later, the reed bed has almost vanished in the wild flower meadow (view from above).

Case study 2 – Oaklands Park, Newnham
Situation/problem
Existing reed bed too small for increased population so new system required needing little maintenance. 55 people using system. Community members on-site throughout day. Large area of second-quality agricultural land available. Slight slope and large clay content.

Solution installed
Aquatron separators followed by waste stabilization ponds with reed beds and willow area at end for disposal.

Notes/reasons
Aquatron installed because compost is a keynote for the community.
- Pond needs little maintenance.
- Large space available.
- Fits in landscape.

Schema/sketch

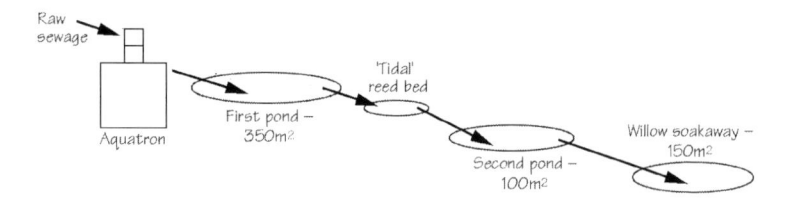

Raw sewage → Aquatron → First pond – 350m² → 'Tidal' reed bed → Second pond – 100m² → Willow soakaway – 150m²

Case Studies 133

First pond under construction (Elemental Solutions).

Ponds always look smaller once the plants have grown. This syetem fits nicely into the landscape.

Case study 3 – Public house, Midlands

Situatlion/problem
Pub and restaurant with plans to expand. Package plant (recirculating biological filter) already existing but failing to meet its consent consistently. Flat field available next to pub garden. Good maintenance in place and grease trap emptied every fortnight. Highly variable loading.

Solution installed
Horizontal flow reed bed – 48 m².

Notes/reasons
Population calculated to discharge at daily rate of 15 litres and 15g BOD per diner, and 12 litres and 12g BOD per drinker, plus 180 litres and 55g BOD per resident. Total of 2,700 litres and 2,075g BOD/day. This is equivalent to 15 resident people by hydraulic load and 20 resident people by organic load. The larger figure (20 people) used, with room to expand. This is over twice what would suffice now but the cost of extra materials was small compared to the costs of organisation and, therefore, a small fraction of the project budget.

Schema/sketch

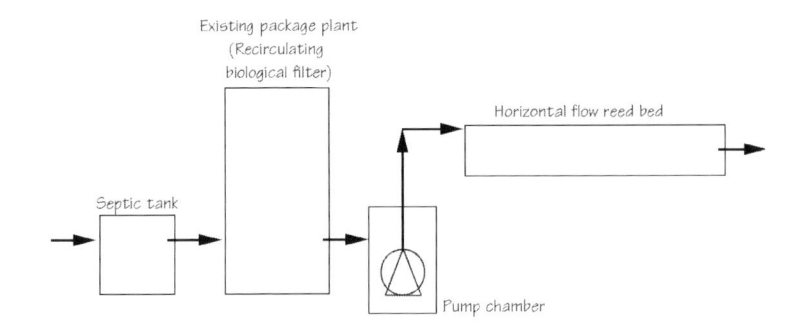

Planting horizontal flow reed bed (Elemental Solutions).

Case study 4 – Wildfowl & Wetlands Trust, Slimbridge, Gloucestershire

Situation/problem

Visitor centre attracting up to 300,000 visitors per year. Existing pumped connection to mains sewerage at end of useful life, and requirement for demonstration of wastewater treatment and wildlife potential of constructed wetlands, as well as on-site sludge treatment. Flat, constricted site, bounded by straight fencing, with requirement to remove nitrogen and phosphorus - i.e. much stricter final effluent standard than required by EA Discharge Consent.

Solution installed

Macerated sewage pumped to primary settlement tank; settled and surface sludge pumped to vertical flow sludge reedbeds (50m^2 times two), settled sewage pumped to multi-stage reedbed comprising:

Vertical flow reedbed – 400m^2
Settlement pond – 50m^2
Horizontal flow aquatic plant bed – 150m^2
Overland flow reedbed – 200m^2
Inundation reedbed – 100m^2

Notes/reasons

Design population 400 resident-equivalents, based on past visitor numbers and occasional peak of 6,000 visitors per day. Sewage treated to Discharge Consent standard (25:45 BOD/SS) by first bed, which also achieves nitrification; subsequent stages for nutrient removal. Horizontal flow bed for denitrification (nitrogen removal), overland flow for suspended solids and BOD polishing, inundated bed of aluminium-rich media, alternately filled and emptied to give good effluent/media contact for phosphate removal. Final (unlined) pond intended for wildlife attraction.

Case Studies 137

Schema/sketch

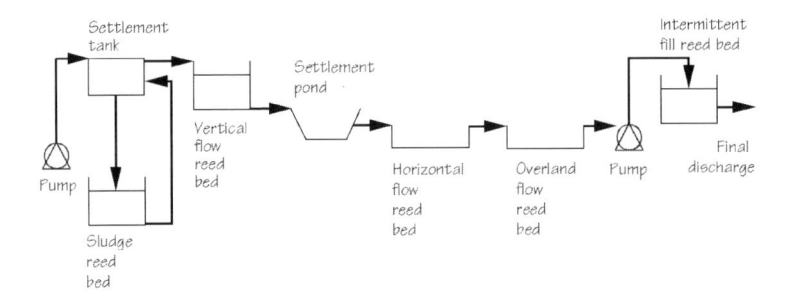

Slimbridge multi-stage aquatic plant treatment system, recently commissioned (Chris Weedon).

Case study 5 – Private house, North London
Situation/problem
Enthusiast with DIY greywater reedbed system behind house. Recycled greywater into house – WCs and washing machine – and even got a rebate on sewage rates! Diverter valves, sensors and computers ran pumps... but the horizontal flow reedbed used was poorly built and the system was not working despite the considerable expense.

Solution installed
Vertical flow reedbeds used with existing reedbed improved. Computer removed, and relays and level-detectors used, so client could be empowered to work with his system.

Notes/reasons
If you know the smell of the trap in the shower you will know why we decided to keep the water moving. So timed recirculation added, to go to garden or back to house. System kept small and simple hence reedbeds. A very costly, energy intensive and high maintenance way to irrigate your garden during hose pipe bans. Main note – don't try this at home folks!

Case Studies 139

Schema/sketch

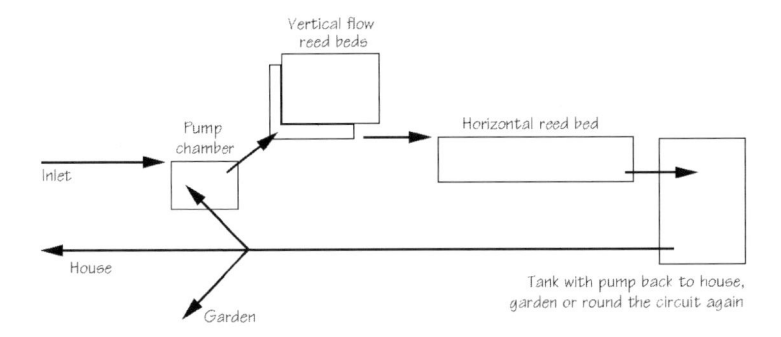

Vertical flow reed beds for greywater re-use system with crystal clear sample (Elemental Solutions).

Case study 6 – Compact Vertical Flow Reed Bed at Will's Barn, author's home – Chris Weedon, Somerset

Situation/problem

Old stone barn purchased for conversion to dwelling when offered separately from nearby house for first time. Sewage from house historically flowed to undersized septic tank and leachfield on land of barn. Leachfield completely blocked, as receiving land very rich in clay. Land slopes steeply away from barn.

Solution installed

Installed sewer from barn to converge with existing sewer from neighbour's house. Raw sewage from both dwellings directed to Aquatron solids separator housed in enlarged existing septic tank. Effluent from Aquatron and from base of composting chamber directed to distribution chamber, feeding a 16m² compact vertical flow reed bed, the effluent of which feeds a small pond. Excellent treatment performance (see *Water Science and Technology*, Vol 48, No.5, pp. 15-23) and provision of habitat for amphibia, dragonflies, etc.

Notes/reasons

High clay content of sloping land of sufficient area and suitable location away from house, argued strongly for gravity-operated reed bed system. Single bed approach developed at this site, avoiding need to swap between parallel beds. Private water supply of limited yield in dry periods, so water conservation essential (achieved mainly by using Ifö 2-4 litre flush toilet (and compost toilet) rainwater harvesting for land irrigation) simultaneously benefiting reed bed performance by decreasing hydraulic load to bed.

Schema/sketch

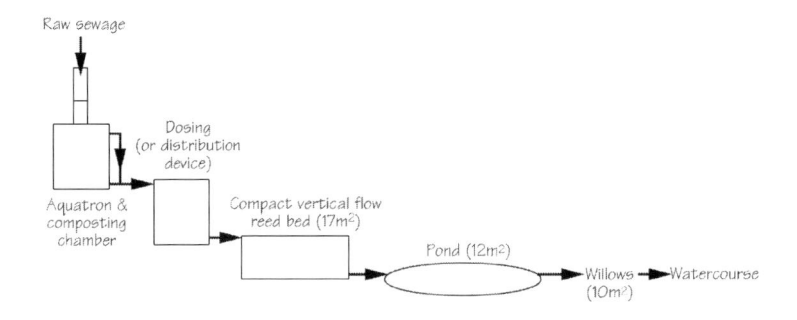

Raw sewage → Aquatron & composting chamber → Dosing (or distribution device) → Compact vertical flow reed bed (17m²) → Pond (12m²) → Willows (10m²) → Watercourse

Compact vertical flow reed bed serving Will's Barn & neighbour (eleventh year) (Chris Weedon).

Case study 7 – Architect's office Herefordshire – Architype

Situation/problem
A rural barn converted to an architectural design studio for around 20 people. The soil is not permeable enough for a conventional soakaway and there is no nearby watercourse to discharge treated effluent. However there is a gentle slope and a reasonable amount of land.

Solution installed
Two chamber horizontal septic tank followed by Flout® dosing device and living soakaway.

Notes/reasons
The site was ideal for a so called living soakaway which utilises the very shallow topsoil to treat and disperse the effluent from an oversize septic tank. The septic tank is of the horizontal two-chamber type rather than a cheaper 'onion' as this is less prone to blockage with extended sludge removal periods. A Flout® dosing device is used to help distribute the settled effluent across the full width of the soil filter.

The system was fitted into the landscape with minimal earthworks but a large volume of woodchip from local tree surgeons. The woodchip helps improve the soil permeability as well as preventing any odour where effluent might sit on the soil surface at the head of the system. The are was planted with common reed but other larger systems have been planted with coppice trees.

Case Studies 143

Living soakaway at architect's office.

Case study 8 – Allotment toilets
Situation/problem
The lack of toilet facilities on allotments creates a barrier to use. Whilst a water supply was available, damage of conventional plumbing would be a serious problem in winter as no power is available for frost protection. Also there was insufficient land for a toilet, septic tank and leachfield. This all indicated that a waterless toilet would be the best solution.

A chemical toilet would need regular maintenance and more compact dry toilets would not be able to cope with the peak load of urine on a busy day. The lack of fall ruled out any compost toilet with a drain in the base unless a pump was to be installed. As wheelchair access was required a more typical vault type compost toilet could not be used due to the flat site.

Solution installed
NatSol Compus 3 Urine separating, twin vault, dry toilet with full wheelchair access.

Notes/reasons
The NatSol toilet system was developed by Nick Grant and Andy Warren to solve all the above problems which are typical of such sites!

Case Studies 145

Case study 9 – Withy Cottage, author's home, Nick Grant, Herefordshire

Nick Grant and Sheila Herring built their energy and water efficient house with local timber, straw and cellulose insulation, but also the judicious use of steel and concrete. Sheila insisted on a compost toilet, having experienced the complete lack of odour compared with a normal toilet. An ES4 WC in the adjacent bathroom provides a back-up for guests and discharges directly into the compost chamber. Subsequently there is no septic tank and the greywater and compost toilet leachate discharge to a Trench Arch leachfield (see leachfields and soakaways, p.56-58).

Other water saving features include taps with the Eco-Plus water saving cartridge, micro-bore hot water pipes to reduce dead legs, an efficient shower and AAA rated washing machine. The result is an indoor water use over the last two years averaging only 51 litres per person per day. This is roughly what is used for toilet flushing alone in a traditional house.

Space heating is by a single woodstove with a small back-boiler and thermal store for hot water and a towel rail. In summer, water heating is by solar panels. Electricity use has averaged 1,533 kWh/year with about half of this to run the office.

Overall, an Eco-minimalist approach of simplicity and attention to detail has allowed them to build to a high specification for around £360/m^2 plus their own labour. Non-water examples of the approach include the lack of central heating (made possible by the insulation) and the fully insulated raft foundation, which is ground and polished to provide a beautiful and practical floor finish. Interestingly the turf roof and straw insulation grab most attention but contribute very little to the building's environmental performance in this situation.

Withy Cottage, a mostly successful attempt at low water and low carbon living without sacrificing comfort.

Perhaps a reaction to seven years washing in a bucket but the luxury is achieved with minimal water and energy use.

Case study 10 – Surface Flow Reed Bed, Herstmonceux Castle, East Sussex

Situation/problem

Owing to large load variation to the sequencing batch reactor, coinciding especially with the coming and going of students, Herstmonceux Castle (university and visitor centre) was experiencing persistently poor quality effluent to the watercourse. Despite much BOD and SS removal by the SBR, discharged effluent often remained above legally consented levels. A large, level, unused area of clay-rich land was available, nearby and downhill of the plant.

Solution installed

Tertiary treatment was required, which was effected by directing the SBR effluent into an unlined 600m^2 surface flow reed bed, operated at 300mm water depth. An unplanted lagoon was installed at the head of the bed, to allow easy occasional removal of solids.

Notes/reasons

Site was also ideal for a subsurface horizontal flow reed bed for which the Castle had obtained two quotes for installation, the lowest of which was more than three times that of the surface flow reed bed.

Case Studies 149

Schema/sketch

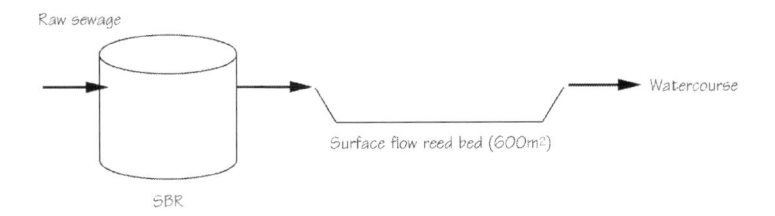

Raw sewage → SBR → Surface flow reed bed (600m²) → Watercourse

Surface Flow Reedbed at Herstmonceux Castle (background) just after planting, with operator Trevor Fox. Note the gravel in the foreground for retention of algae, etc.

Case study 11 – Prospect House, author's home, Mark Moodie, Gloucestershire

In 2003 we obtained a barely improved miner's cottage in the Forest of Dean with great views through the rampant brambles. There were very poorly laid drains around the house but the main sewer was available. There was no insulation in the dark, cold house apart from a duvet loosely wrapped around the copper water cylinder.

First we re-laid the drains and secured our mains connection. When some walls had been moved we focused on best practice water use and energy efficiency. Pipe runs were optimised, sub-water-meters installed, flow regulators put on taps and shower, and water-efficient white goods were purchased; washing machine (at last!) from the Which? guide and Ifö WCs (one siphon flush ES4 and one valve flushed, close-coupled version for comparative evaluation). Now we just need to keep reminding the kids about taps. Because the main sewer was available (see The Designer's Flowchart in chapter 5) we were pleased to leave our muck to the Authority. This means we have the time to concentrate on other aspects of our ecological responsibility.

Our main mental, physical and financial input has disappeared into insulation (walls, floors and ceilings) and heating: a wood pellet boiler, solar water system, stratifying high-capacity super-insulated solar thermal store for mains pressure supply, underfloor heating, and UK-timber high specification double glazing.

I think we have done well. It takes three bags of wood to heat the house on a cold day, there are no radiators eating into the narrow rooms and the light is lovely. Water use for two children, one just coming out of (washable) nappies, and Beki and I is 45.1 litres each per day. Electricity use for the house and construction is about 1,800 kWh/year.

Case Studies 151

Author Mark Moodie's home, Prospect House.

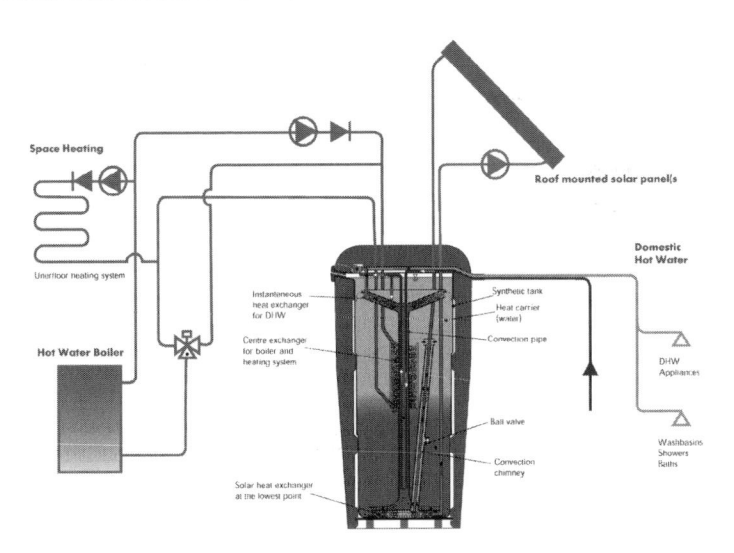

Solar thermal storage tank to provide mains pressure hot water on demand and space heating support.

Case study 12 – Upton Bishop Church, Herefordshire

Situation/problem

Many rural churches are now providing toilet facilities. But where should the sewage go if there is no sewer and the church is surrounded by graves? Case study 3 shows one solution but in this case there was no basement.

Solution installed

Trench arch disposal and treatment. This is a carefully designed long wide-bore chamber, made from tough pipe. Unsettled sewage is discharged in one end and is spread out in the solids treatment section for aerobic decomposition and worm degradation on the base. Water carries on to the second section where it percolates into the soil.

Notes/reasons

The trench arch is ideal because of intermittent use, allowing periods of soil recovery. The system needs only a 400mm deep excavation and there is no septic tank. The Trench Arch was developed by Elemental Solutions. The process works well in this lightly loaded situation. For domestic installations such a system is not necessarily sufficient, due to higher and more consistent loading although long-term trials are underway. Each installation, especially an unusual one such as this, needs to be examined and agreed upon by the Environment Agency.

For more information, see the Elemental Solutions website: www.elementalsolutions.co.uk/examples

Case Studies 153

Schema/sketch

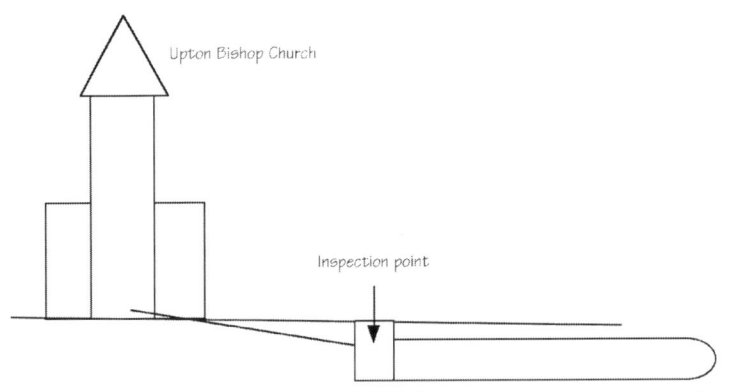

Upton Bishop Church

Inspection point

Trench Arch in graveyard takes waste water directly

Exposed trench arch.

Case Study 13 – Bodiam Castle, East Sussex: National Trust Visitor Centre

Situation/problem
Major NT visitor attraction plus pub and cottages, experiencing large, sudden variations in sewage load (visitor numbers), at a flat site, with high water table in a flood zone. Sole option for siting was next to site entrance and neighbouring gardens (potential odour issues) to avoid area of Scheduled Ancient Monument. Two previous mechanical treatment plants had performed unreliably, eventually resulting in serious Discharge Consent failure and a fine from the Environment Agency. Connection to the mains was about to proceed at high capital cost.

Solution installed
On-site treatment at one-tenth cost of mains connection, using slightly under-sized mechanical plant followed by vertical flow reed bed. RBC was selected as the mechanical treatment stage, mainly owing to its on-going power consumption being as little as one tenth of equivalent alternatives; particularly important at visitor centres, where there is low use for long periods. The RBC effluent was pumped up to a distribution chamber, providing gravity flushing at 40 l/s to the surface of a lined 208m² compact vertical flow reed bed. The bed was raised slightly above ground level for flood protection, which also allowed gravity recirculation of the fully nitrified effluent to the RBC inlet.

Schema/sketch

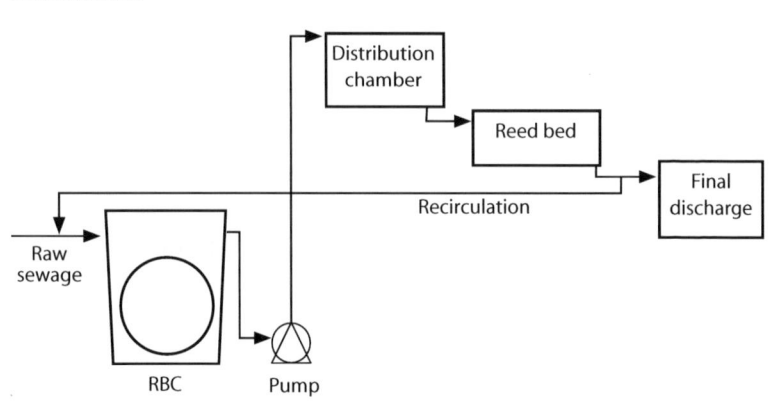

Notes/reasons

Using the two technologies in series plays to their strengths, while cancelling out limitations: mechanical plant cannot reliably treat sudden large load fluctuations, but produces "zero"-odour effluent for distribution to surface of reed bed; reed bed reliably treats primary settled effluent but this would involve foul odour release during intermittent dosing to bed surface, avoided by pre-treatment by mechanical plant, which also removes most of the BOD, allowing complete nitrification by reed bed. To summarise the overall rationale, the approach provides:
- Robustness to sudden load fluctuations – because of the extensiveness of reed bed.
- Odour avoidance – by adding dissolved & combined oxygen (nitrate) to RBC inlet.
- Nitrogen removal – by denitrification in RBC primary settlement stage.

RBC (not shown) plus compact vertical flow reed bed (9 months old in photo) serving cottages, pub (background) and NT visitor centre.

Glossary

Adsorption
When one substance is taken up by another adhering it to its surface.

Aerobic
Containing free oxygen (i.e. oxygen gas). Usually refers to conditions required for certain types of bacteria (aerobes) to thrive. Typically, aerobic processes require at least 1 mg/l dissolved oxygen.

Algae
A collection of relatively simple plants, varying in size from single cells to filamentous forms, such as 'blanket weed', to several meters in the case of some seaweeds.

Algal blooms
Extensive growths of algae indicating eutrophication. Can be a problem in sewage systems using ponds, as algae discharged in effluent will give a high SS and BOD reading.

Ammonia (NH3)
A sharp smelling gas that readily dissolves in water, forming relatively non-toxic ammonium ions (NH4+). Poisonous to fish and oxygen-consuming when metabolised.

Anaerobes
Organisms capable of living in the absence of free oxygen. Strict anaerobes cannot survive in the presence of oxygen.

Anaerobic
Containing no free oxygen. Anaerobic sewage smells unpleasant.

Anoxic
Without free oxygen but with chemically-combined oxygen such as nitrate (NO3-) present, which can be used by certain bacteria as a source of oxygen, so releasing nitrogen gas.

ATU
1-allyl-2-thiourea, added during the BOD test to inhibit nitrifying bacteria, so that the oxygen demand measured is due solely to the breakdown of carbonaceous (organic) matter.

Bauxite
Naturally occurring aluminium oxide, used in the production of zeolite for phosphate-free washing powders.

Blackwater
Domestic wastewater from toilet flushings, as opposed to greywater (qv.).

BOD
Biochemical oxygen demand, amount of oxygen (mg/l) leaving a water sample of known volume during a five-day (usually) incubation at 20°C, in the dark. Indicates concentration of rapidly biodegradable organic (carbon-containing) material in the sample.

Chloride (Cl-)
Humans excrete around 6g of chloride a day in sewage. Since treatment plants do not generally remove it, its concentration is a useful indicator of sewage pollution in a watercourse. In normal concentrations chloride is not considered harmful.

Chlorine (Cl2)
Toxic yellow-green gas used to sterilise drinking water (and used in chemical warfare). The element chlorine (Cl) is found in table salt, as well as in some of the most toxic, carcinogenic and persistent chemicals made by humans, such as DDT and Dieldrin. Chlorine is responsible for huge reductions in water-related illness and is implicated in the possible increase in potential carcinogens in our water.

Detergents
Products used for cleaning in solution with water, usually with the help of surfactants. Unlike soaps they are effective in hard water and tend to be less biodegradable.

Digestion
Chemical or biological process in which compounds or materials are broken down into simpler compounds.

Dip pipe
'T'-shaped (on its side) pipe that allows liquid from below the surface to leave a tank while retaining the surface material.

Dissolved oxygen (DO)
Free oxygen dissolved in water is required for higher plants, animals and aerobic bacteria, and so a measure of the dissolved oxygen in a stream or pond can provide an indication of its health. DO is usually expressed as mg/l or as a percentage of saturation at the sample's temperature.

Effluent
Wastewater leaving a treatment system or stage of a system (cf. influent).
Enzymes
These proteins are produced by living cells to act as catalysts. They increase the rate of chemical reactions (such as those involved in decomposition), yet remain themselves unchanged. There are products on the market containing dried enzymes for adding to septic tanks or blocked drains, which can accelerate organic decomposition. Biological washing powders contain enzymes, which are added to help digest proteins but these types of enzymes can lead to health problems in sensitive people.
Eutrophication
Enrichment of a body of water by nutrients, usually nitrates and phosphates. This can lead to excessive algal and plant growth. Phosphate is usually the limiting nutrient in fresh water and unfortunately is the hardest to remove from wastewater. Eutrophic waters lack species diversity and tend to be dominated by a few species of algae.
Facultative bacteria
Bacteria able to survive in either aerobic or anaerobic conditions.
Greywater
Domestic wastewater not including toilet flushings, as opposed to blackwater (qv.).
Infiltration
In geological terms this is the vertical flow of water from the soil surface towards the water table. In sewage circles it usually refers to water entering a sewer from the ground, e.g. through broken pipes and manhole walls. Also used to describe a method of disposal to land, such as a soakaway.
Influent
Wastewater entering a treatment system or a stage of a system (cf. effluent).
Metabolism
The transformation of matter and energy in organisms. Subdivided into catabolism (qv.) — the breakdown of complex organic molecules into simpler products, and anabolism (qv.) — the synthesis of complex molecules from simpler ones.
Night soil
Undiluted human muck, so called because it used to be collected at night from earth closets or out-houses.

Nitrate (NO_3^-)
One of the three major plant nutrients, non toxic to aquatic life (in normal amounts) but capable of leading to eutrophication. A maximum level is set for nitrate in drinking water as it can cause 'blue baby syndrome' and it may be implicated in stomach cancer since it can form nitrite in the stomach.

Nitrite (NO_2^-)
A nitrogenous ionic molecule poisonous to fish (see ammonia and nitrate). Nitrite is highly unstable and is usually formed when nitrate is being created or broken down (eg during denitrification). Although measured in effluent samples, its concentration is of limited interest and rarely rises above 1mg/l in wastewater because of its instability.

Nutrient removal
The removal of phosphorus and nitrogen compounds in effluent to reduce the risk of eutrophication. A ratio of BOD to nitrogen and phosphorus of about 100:5:1 is required in order to optimise aerobic digestion of wastewater. Domestic sewage has a BOD:N:P ratio around 100:17:5. This means that aerobic digestion will leave an excess of N and P, which may, therefore, escape to a watercourse. Nitrogen can be removed by denitrification but phosphorus has no significant gaseous pathway and so must be removed in biomass, by adsorption or chemical precipitation.

Phosphorus (P)
Total phosphorus in domestic sewage amounts to around 2.8g/person/day of which about half is from washing powder. In eutrophication, phosphorus is usually the limiting nutrient. Algal blooms usually occur at phosphate levels greater than 0.05 mg/l in still water and 0.1 mg/l in running water. Typical sewage effluent is in the range 5-15 mg/l.

Population equivalent
Specifies the loading from places of intermittent use such as restaurants or hospitals in terms of an equivalent full-time domestic population. It is assumed that one person equivalent represents about 60g /day of BOD and about 180 l/day hydraulic loading, 10g nitrogen/day and 2.8g phosphorus.

Retention time
A measure of the average time taken for wastewater to pass through a treatment unit such as a pond, tank or reed bed. The theoretical retention time is obtained by dividing the tank volume in litres by the daily flow in litres per day. In practice, short-circuiting reduces the actual retention time to around half of this for most ponds and settlement tanks.

SS
Suspended solids, concentration of particulate material (mg/l) removed via a fine filter from a water sample of known volume.

Soap
Material used for washing; a mixture of sodium salts of stearic, palmitic or oleic acids or of the potassium salts of these acids ('soft soap'). Soaps are made by the action of sodium hydroxide (caustic soda) on fats. This is why weak caustic soda solution feels soapy; it turns your greasy finger tips into soap!

Surfactant
Short for 'surface active agent', used in conjunction with detergents as a wetting agent.

Resources

Consultancy
Elemental Solutions, Nick Grant, tel: 01981 540728,
www.elementalsolutions.co.uk
Watercourse Systems Ltd, Chris Weedon, Will's Barn,
Chipstable, Taunton, Somerset, TA4 2PX; tel. 01984 629070,
mobile. 07801 924493, email: weedon@compuserve.com
Centre for Alternative Technology, Pantperthog, Machynlleth, Powys,
SY20 9AZ; tel. 01654 705950, fax. 01654 702782,
email. consultancy@cat.org.uk website www.cat.org.uk

Courses
Centre for Alternative Technology (as above).
Chartered Institution of Water and Environmental Management,
15 John Street, London, WC1N 2EB; tel. 020 7831 3110,
fax. 020 7405 4967.
Institute of Irrigation and Development Studies,
University of Southampton, Southampton, Hants, SO17 1BJ;
tel. 01703 593728, fax. 01703 677519.
Silsoe Research Institute, Cranfield University, Wrest Park, Silsoe,
Bedford, MK45 4HS; tel. 01525 860000, fax. 01525 860156.
University of Abertay, Waste Water Department, Bell Street, Dundee,
Scotland, DD1 1HG; tel. 01382 308000, fax. 01382 308877.
Water Research Centre, plc, Henley Road, Medmenham, Marlow,
Bucks, SL7 2HD; tel. 01491 571531, fax. 01491 579094.
Watercourse Systems Ltd., (as above).

Index

Acinetobacter 63
Activated Sludge Package Plant 33, 35
Aerobic 15-16, 27, 32, 34, 36-37, 39, 42, 62-64, 68, 152, 157-160
Algae 99, 121-123, 149,- 157, 159
Aluminium 63, 98, 121-122, 136, 158
Ammonia 9, 14-16, 18-19, 39, 41, 59-62, 66-68, 74, 80, 83, 121, 157, 160
Anabolism 10-11, 159
Anaerobic 15,-16, 26, 42, 63-64, 83, 157, 159
Aphelocheiridae 69-70
Aquaculture 119, 122
Ascaris lumbricoides 83
Ascaris worms 83
Astacidae 69-70
ATU (allylthiourea) 57, 60, 75, 157
Autotrophs 10-11
Bacteria 4, 10-11, 14-15, 18, 28, 41, 59-62, 64-65, 74, 83, 99, 125, 157- 159
Basin 105, 109-110, 112, 115
Bath 25, 39, 105, 109, 110, 112, 114
Bauxite 101, 158
Big Circle 7, 10-13, 58, 61, 77, 89
Biodegradability 98-99
Biofilters 28, 34

Biological filters 28, 32
Biotic index 69-70
Blackwater 77, 158-159
Bleach 98-99
BOD 15, 17-19, 39, 51, 54, 55-62, 65, 67-68, 74-76, 121, 123, 134, 136, 148, 155, 157-158, 160
Building Regulations 24, 37, 39-40, 73, 102
Building sites 89
Caravans 89
Carbon 10-11, 16, 58-60, 62, 65-66, 78, 80, 85-86, 89, 93, 147, 158
carbon breakdown 59
carbon dioxide 10, 16, 58, 78
Catabolism 9-10, 159
Cattle slurry 61
Cesspools 22-24
Cesspools (containment) 23
Chemical oxygen demand 60
Chemical toilets 88
Chemotrophs 11
Chironomidae 69-70
Chloride 14, 99, 158
Chlorine 99, 158
Cistern 103, 106-108, 110
Clinker beds 28
Clivus Multrum 81, 83
Cobalt 58
COD 60

Colloids 79
Compost 80
Compost toilet 3, 80, 82-83, 90, 96, 96, 107, 140, 144, 146
Cray fish 70
Dead legs 105, 110, 146
Decay 7, 123
Denitrification 61-62, 67, 121, 136, 155, 160
Detergents 2, 7, 14, 98-100, 102, 114, 158, 161
Determinands 55-56, 59, 61-62, 64-66, 71, 76
Digestion 9-10, 16, 57, 158, 160
Discharge 19, 23, 25-26, 28, 32-34, 37-38, 40, 45-46, 48, 50, 55, 57, 60-61, 63, 68, 70-71, 73-74, 98, 119-124, 134, 142, 146, 154
Discharge Consent 136, 154
Disease 4, 64, 82
Dishwasher 99, 115
DIY 37, 39, 44, 48, 75, 105, 111, 113, 124, 138
DOWMUS 80
Dry toilets 23, 93, 97, 106, 144
Dual flush 106-108, 111
Ecological sensitivity 22
Electric toilets 88, 91
Environment Agency 60, 72-73, 75, 101, 121, 152, 154
Environmental health 1, 48
Environmental Health Department 74, 75
Environmental Quality Index 72
Environmental regulators 50, 72-73, 76
Escherichia coli 64
Eutrophication 63, 98-99, 157, 159- 160
Faeces 4, 7, 64, 82, 85, 89, 93, 96
Farmyard washings 61
Fats 14, 28
Fatty acids 14, 16
Fertilizer 93
Filter beds 28
Fish 11, 13, 15, 59, 68, 70, 157, 160
Flies 82, 89, 93
Flowforms 66
Flow regulator 109, 112, 150
Free-draining 39, 77
Gas 10, 15-16, 58, 60, 62, 67, 79, 157-158
Gravel 36, 39, 41-42, 44, 67, 149
Greenhouse 49, 116
Greywater 23, 27, 99, 103, 114, 117, 138-139, 146, 158
Gypsum 114
Health 1, 44, 48, 56, 65, 68, 74-75, 82-83, 90, 115, 158-159
Heavy metal 41
Heterotrophs 10-11
Hills, Lawrence 78, 125-126
Horizontal flow reed bed 39-40, 135, 148
Household solvents 101
Humanure 4, 77, 80, 82-83, 85, 90, 93, 96, 120
Humus 2, 28, 30, 51, 77-80, 85, 96, 119-120, 123, 130
Hydrogen 16, 58, 64, 79
Hydrogen sulphide 16, 64
Hydroponic 121
Hygiene 78, 80, 90, 93
Indicators of water quality 70
Irrigating 119, 124-125
Irrigation 23, 46, 50, 99, 101, 121, 140

Jam-jar monitoring method 19
Jenkins, Joseph 85
Lagoons 42, 123
Leachfield 22-23, 25-26, 44-46, 49, 97, 98, 102, 116, 120, 140, 144, 146
Leach pit 44
Leaks 23, 44, 52, 103, 105
Lignins 79
Living determinands 71
Living machines 23, 48
Living Machines 48
Low flush toilets 107
Mains 3, 4, 8, 21, 73, 89-90, 109-110, 112, 114-115, 136, 150-151, 154
Maintenance 22, 28, 30, 34, 40, 42, 46, 106, 132, 134, 138, 144
Methane 16
Microorganisms 8-10, 13, 15, 17-18, 28, 32-33, 41, 56-61, 65, 116, 119, 121
Midge larvae 70
Milk 57, 61
Mineralisation 9
Minerals 7-8, 10, 14, 48, 79, 83, 120-121, 123
Mineral solution 8-10, 16, 18-19, 21, 61, 79, 119-120
Monitoring 53-54, 69-70, 119, 121
National Sanitation Foundation 83
Nitrate 62, 160
Nitrite 62, 66, 160
Nitrobacter 60
Nitrogen 59, 89, 155, 160
Nitrosomonas 60
NRA 60, 121
Nutrient 119, 160

Nutrient solution 10, 120
Oak 125
Ocean Arks International 48
Odour 27, 30, 41, 44, 46, 50, 64, 74, 82, 106, 122, 142, 146, 154-155
'off-the-shelf' treatment systems 30
Oligochaeta 69-70
Outflow 65
Outlet 23-24, 27, 51, 59, 65, 67
Overflow 25, 45, 107, 123
Oxygen 9, 13-16, 18, 32-33, 39, 43, 56-64, 70, 77-78, 155, 157-158
Oxygenation 41, 63
Oxygen removed 57
Package plants 23, 28, 30, 32-35, 129
Pathogen 4, 35, 65, 82-83
Pathogenic organisms 19, 64, 83
Pathogens 4, 55, 59, 64-65, 83, 85, 90, 93, 121
Percolating filters 23, 28, 32-33, 36
Phosphate 14, 16, 63, 66-68, 98-99, 101, 136, 158-160
Phosphorus 16, 19, 55, 58-59, 63-64, 66-67, 74, 86, 89, 93, 121-122, 136, 160
Phototrophs 11
Phragmites 36, 42
Pollution 23, 56, 61, 70, 74, 98-99, 101, 121, 123, 127, 158
Poplar 125
Preliminary treatment 17
Primary solids 56
Protozoa 65
Pumps 17, 32, 35, 103, 110, 138
Rainwater 51, 101-103, 114-115, 140

Rainwater harvesting 114, 140
Receiving watercourse 23
Recirculating biological filter 32-33, 134
Reed beds 40-43, 62, 72-73, 78, 94, 102, 106, 114, 125, 130, 132, 139
reed beds horizontal flow 23, 39, 40, 42, 94
Regulations 23-24, 28, 36, 39
Regulator 70, 72-74, 112
Regulators 28, 50, 54, 72, 74, 76, 109, 150
Reusing Water 53
River bugs 70
River Invertebrate Prediction and Classification System 72
Rodents 82
Rotary biological contactor 31-32
Rotating arm systems 28
Royal Commission Standard 74
Sand 9, 36-37, 39, 41, 67
Secondary treatment 17-18, 20, 25, 28, 30, 39, 42, 46
Seidel, Käthe 125
Septic tank 1, 3, 17, 19, 22, 24-28, 39, 44, 49, 65-66, 83, 98, 116, 130, 140, 142, 144, 146, 152
Settlement ponds 42
Settlement tanks 24, 52, 102, 160
Sewage 1, 3, 4-5, 7-25, 28, 30, 32-33, 35, 36, 38, 39, 41-42, 45-46, 48, 51-66, 68, 70-71, 73-79, 83-84, 88, 90, 96-99, 101-102, 105, 116-117, 119-127, 129, 136, 138, 140, 152, 154, 157-160
Sewage ponds 42
Shower 104, 109-110, 112, 138, 146, 150
Showers 105, 109-110
Silage effluent 61
Sinks 14, 110
Siphon flush 108, 150
Sludge 1, 10, 16, 23-25, 27-28, 32-35, 44-45, 52, 63-64, 83-84, 90, 96, 136, 142
Smell 4, 17, 22, 54, 64, 114, 122, 138
Soak 28, 80, 82-83, 87, 90, 102
Soakaways 23, 44-45, 49, 146
Soap 114, 161
Soil 39, 44, 46, 48-49, 63, 77-80, 82, 88, 90, 96, 101-102, 114, 119-126, 129, 142, 152, 159
Space 17, 37, 39, 79, 84, 98, 102, 132, 151
Strongyloides stercoralis 83
Submerged biological aerated fixed film system (BAFF) 33
Sulphide 14, 16, 64
Surface flow reed bed 23, 42-43, 148
Taboo 11, 83, 93, 122
Taps 52, 103, 105, 110-112, 146, 150
Tariff 113
Temperature 9, 15-16, 57, 60, 65, 82-83, 100, 109-110, 112, 123, 158
Tertiary treatment 19-20, 39, 41, 42, 44, 46, 48, 120, 123, 148
TKN 62, 67
TOC 60, 65, 68
TOC (total organic carbon) 60, 65
Todd, John 48
Topsoil 46, 49, 78-79, 90, 142

Trickling filters 28
Turbidity 54
Urea 14, 59, 61
Urine 4, 7, 14, 82-83, 85-87, 89-91, 93, 96, 144
Urine-separating toilet 86
Urtica dioica 126
Valve flush 111, 150
Vegetable washings 61
Vertical flow reed bed 23, 28, 36, 39, 78, 140, 154
Washing machine 68, 100, 115, 138, 146, 150
Washing powders 63, 66, 98-99, 158-159
Water audit 104
Water conservation 23, 97, 102-104, 113, 116-117, 140
Water cycle 116
Water meters 105, 113
Water recycling 115-116
Wildlife 42, 44, 48, 125, 130, 136
Willows 23, 46, 119, 124-125
Worms 15, 69-70, 79, 83
Zeolite 98-101, 158
Zinc 58
Zoogloea 28